Numerical C

Applied Computational Programming with Case Studies

Philip Joyce

Apress®

Numerical C: Applied Computational Programming with Case Studies

Philip Joyce
Goostrey, UK

ISBN-13 (pbk): 978-1-4842-5063-1 ISBN-13 (electronic): 978-1-4842-5064-8
https://doi.org/10.1007/978-1-4842-5064-8

Managing Director, Apress Media LLC: Welmoed Spahr
Acquisitions Editor: Steve Anglin
Development Editor: Matthew Moodie
Coordinating Editor: Mark Powers

Cover designed by eStudioCalamar

Cover image designed by Freepik (www.freepik.com)

Distributed to the book trade worldwide by Springer Science+Business Media New York, 233 Spring Street, 6th Floor, New York, NY 10013. Phone 1-800-SPRINGER, fax (201) 348-4505, e-mail orders-ny@springer-sbm.com, or visit www.springeronline.com. Apress Media, LLC is a California LLC and the sole member (owner) is Springer Science + Business Media Finance Inc (SSBM Finance Inc). SSBM Finance Inc is a **Delaware** corporation.

For information on translations, please e-mail editorial@apress.com; for reprint, paperback, or audio rights, please email bookpermissions@springernature.com.

Apress titles may be purchased in bulk for academic, corporate, or promotional use. eBook versions and licenses are also available for most titles. For more information, reference our Print and eBook Bulk Sales web page at http://www.apress.com/bulk-sales.

Any source code or other supplementary material referenced by the author in this book is available to readers on GitHub via the book's product page, located at www.apress.com/9781484250631. For more detailed information, please visit http://www.apress.com/source-code.

Printed on acid-free paper

Table of Contents

About the Author

Philip Joyce has 28 years of experience as a software engineer – working on control of steel production, control of oil refineries, communications software (pre-Internet), office products (server software), and computer control of airports. Programming in Assembler, COBOL, Coral66, C and C++. Mentor to new graduates in the company. He received his MSc in Computational Physics (including augmented matrix techniques and Monte Carlo techniques using Fortran) from Salford University in 1996. He is also a chartered physicist, a member of the Institute of Physics (member of the Higher Education Group), and has been a teacher of mathematics to 11–18-year-old students for 14 years.

About the Technical Reviewer

 **Michael Thomas** has worked in software development for over 20 years as an individual contributor, team lead, program manager, and vice president of engineering. Michael has over 10 years of experience working with mobile devices. His current focus is in the medical sector using mobile devices to accelerate information transfer between patients and health-care providers.

Acknowledgments

Thanks to my wife, Anne, for her support, my son Michael, and my daughter Katharine. Michael uses regression techniques in his work and has shared some ideas with me. Katharine was the catalyst for me writing the book. While she was at university, I taught her Fortran programming with applications in the mathematics she had done in her first year. The work we did was based on my MSc course in Computational Physics but tailored toward mathematics applications. This work then became the basis for this book.

Thanks to everyone on the Apress team who helped me with the publication of this, my first book. Special thanks to Mark Powers, the coordinating editor, for his advice; Steve Anglin, the acquisitions editor; Matthew Moodie, the development editor; and Michael Thomas, the technical reviewer.

Introduction

This book is about learning to write computer programs in C to solve problems in mathematics and to show how C can be used to get data information in other areas like economics and biochemistry.

Computers are used in most areas of life, and it can be vital to your area of work to realize how helpful computer software can be. This book aims to show you how you can write software in C to help you in your particular field.

The book uses your existing knowledge of basic mathematics to demonstrate to yourself how useful software can be. Many mathematical problems are solved using logical algebraic techniques which are taught at school. These are done with pen and paper and form the foundation of mathematics. However, there are a lot of problems which would take a group of people working together hours to solve in this way but could be solved in seconds using computer software. In fact, some problems can only be solved using computer software.

The book starts by introducing the C language and showing you how to write a simple program very quickly. The chapters take you through mathematics you probably already know from school. You will have solved problems in algebra using pen and paper. One topic is "Trial and Improvement" whereby you take an equation that cannot be solved using normal analytical methods, but you can try any value out in the equation and see if you get the value at the right of the equation. As an example, we can substitute any value of x into the following equation and see if we get 13.

$$5x^4 + 17x^3 - 3x = 13$$

If your answer is higher than 13, you try a lower value of x to plug into the equation. If this gives a value lower than 13, then you know that the correct value of x must be somewhere in between your first value of x and your second value. So using this method, you can just keep narrowing down closer and closer to the correct right-hand value. This is a perfect problem to be solved by writing a C program. Whereas the problem could take you half an hour to solve using pen, paper, and a calculator (depending on the complexity of the equation), a C program will do it in seconds.

Another problem you may have done at school is using the quadratic formula to solve quadratic equations. (Remember that a quadratic equation is one where the highest power of x is x^2, e.g., $2x^2 + 3x - 5 = 0$.) These are solved using a formula which, at first sight, might look a bit daunting.

$$x = \frac{-b \pm \sqrt{b^2 - 4ac}}{2a}$$

But this is just the solution to the quadratic equation

$$ax^2 + bx + c = 0$$

where in our equation earlier, a is 2, b is 3, and c is –5.

We can just substitute these values into the formula in our C program and get the solution to the equation.

We can also write software to solve simultaneous equations. In calculus we can write a quick and efficient way of using the Trapezium Method of integration. Another method of integration has the exotic name "Monte Carlo Integration" in which the C program makes use of the way it can generate random numbers. Other mathematical methods include Simpson's Rule, matrix arithmetic, regression, the Product Moment Correlation Coefficient (sounds impressive), another Monte Carlo method (this time to find pi), the augmented matrix method for solving simultaneous equations, and the solution of differential equations.

Finally, we look at methods of using C to access and retrieve information from data files. These methods can have applications in many fields, but in this book we will look at simple examples in medicine and economics.

The book can be used in educational organizations and commercial organizations or can be used by individuals who wish to further their knowledge.

If you don't have the C program development tools, you can download them free of charge from Microsoft and from other organizations.

You can also download software packages to draw graphs. One of these is Graph which is free to download. In Graph you can just enter an equation and it draws the curve. You can also insert a set of data points, from a file that your program creates, into the Graph package and it will display your points. You can then see from the pattern of the points if it follows the shape of a known function.

CHAPTER 1

Introduction to C

The C programming language was created in the 1970s, yet it is still in extensive use today and is the basis of many other languages. For this reason I have used C as the language for the solution of the numerical problems demonstrated in this book.

The level of C used will be sufficient to solve the numerical problems here, but an appendix is included at the end of the book to show extensions that you may want to use in your solutions here or in your work with C in the future.

If you don't already have a C development environment on your computer, you can download it, free of charge, from Microsoft. You can load their Microsoft Software Development Kit (SDK). Another way you can access C is by using Visual Studio. Again, a version of this can be downloaded.

Note   Appendix A contains a guide to two development environments.

First Program

What is generally regarded as a good introduction to a programming language is writing code that just prints a simple message to the screen.

The following is an example.

```
/* This is my first program in C */
int main()
{
        printf("My first program\n");
        return(0);
}
```

© Philip Joyce 2019
P. Joyce, *Numerical C*, https://doi.org/10.1007/978-1-4842-5064-8_1

The "\n" at the end returns the cursor to a new line. Try this for yourself. Open notepad. Type in the code and then save the file to the directory where you want to save your programs. When you save it, put ".c" after the name (i.e., if you want to call it myfirstprog, then call it myfirstprog.c).

Now you have to compile it. Do this by typing "`cl myfirstprog.c`". Compiling converts your written code into "machine code" which the hardware in the computer understands. It also links in any other software that your program might need.

`int main()` delimits your code between the { and the } (although we will see later that you can write a separate piece of code outside of the `main()` part and call it from the `main()` part.

"`printf`" in your code tells the computer to print whatever is between each of the double quotes. `return(0);` indicates that you detected no errors while calling `printf`. Make sure you put the semicolon ; after the statement. This tells the compiler that it is the end of the instruction. If you don't do this, the compiler will take anything following this (in this case, the `return(0)`) and assume it is part of the same instruction. This will cause the program to fail.

It is good practice to give your program a name that describes what it does so that when you list all of your programs in your directory, you will know which one to look at. It is also important to put a comment at the start of each program to say what it does. This is `/* This is my first program in C*/` in the code. Comments are usually also written within your program to describe what a slice of code does or even what a single line of code does. The compiler ignores everything inside /* and */. BUT BE CAREFUL. If you forget to put the */ at the end of your comment, the compiler will think everything following is a comment. Try this for yourself. Change your code for myfirstprog to take out the end of comment marker (*/). Then compile it and run it. You should get a "`fatal error`".

Get and Print a Character

Now that we can display a message to the person running our program, we can ask them to type in a character, then read the character, and print it to the screen. One way we do this is by using the instructions `getchar` and `putchar`. Here is an example of code to do this.

```
#include <stdio.h>
/* read and display a number */
int main () {
    char c;

    printf("Enter character: ");
    c = getchar(); /* read the character in */

    printf("Character entered: ");
    putchar(c); /* write the character */

    return(0);
}
```

int main(), printf, and return are similar to those in your first program. char c; means that you are reserving a place in your program where you will store the character which is read in. c can then be referred to as a variable in your program. getchar() reads the character that the person running your program has typed in, then putchar() prints the character back to the screen. In the code c=getchar() the = sign means "assign to." So the instruction is saying get the character and assign it to the variable c.

Note   A variable is just an area of the computer's memory that we want to use. We give these areas names so that we can access them easily.

Later we will see code where we want to know if one variable (say x) equals another (say y). In this case we have "==" to mean equals, for example, "if x == y". Try typing in the preceding program. Call the program anything you like but, again, something that helps you remember what the program does is good practice. Compile your program using cl progname.c then run it by typing in progname. Try typing in a character. Your program should reply with the character you typed in. Try typing in your first name. What happens? getchar() only reads one character and it will only store the first character you typed into the char c data store in your program. #include<stdio.h> is a command to tell the compiler to attach to your executable program the code which executes the getchar() and putchar(). stdio refers to the standard input and output library. Note the comments in the program telling you what is going on. This is only for your benefit when reading through your program in the future. It can be omitted and the program will run exactly the same.

Add Two Numbers

Now that we know how to prompt the user to enter a number, we can extend our program to let them enter two numbers, then, in our program, we add them and display the answer.

Here is some code to do this.

```c
/* Read in two integers , add  them  and display the answer */

#define _CRT_SECURE_NO_WARNINGS
#include<stdio.h>

int main()
{
    int this_is_a_number1, this_is_a_number2, total;

    printf("Please enter an integer number:\n ");
    scanf("%d", &this_is_a_number1); /* read number in */
    printf("You entered %d\n", this_is_a_number1);

    printf("Please enter another number: \n");
    scanf("%d", &this_is_a_number2); /* read number in */
    printf("You entered %d\n", this_is_a_number2);

    total = this_is_a_number1 + this_is_a_number2;/* add two numbers */
    printf("total is %d\n", total);

    return 0;
}
```

In this program we are reading in integer numbers that can be up to 10 digits. We define the storage for each of our numbers using int as shown at the start of the program. We have also specified storage for where we want to store the total when we have added our numbers. This is total. Notice that we can list all our storage names next to each other after the int command, as long as they are all int types.

Note   Types are the way we differentiate between our data, for example, whole numbers are integer or `int` and characters such as "A", "$", and "?" are `char` types. More information about types of data can be found in Appendix B.

In this program we use `scanf` to read the characters from the screen rather than `getchar()`. This is because our numbers to be added can be more than one character. The `%d` in `scanf` and `printf` specifies an integer to be read or written. In `printf` here the answer to be printed is stored in "`total`".

Type in the code and, again, give your program a meaningful name. Try entering numbers. Check your answers.

Enter positive and negative numbers. Check your answer.

Note that your numbers must be integers (whole numbers).

(BUT WHAT IF WE WANT TO ADD DECIMAL NUMBERS?)

Add Two Decimal Numbers

This code is similar to the code that added two integer numbers. In this case we can add decimal numbers. We define the storage for these as `float`, meaning floating point.

```
/*  Add two floating point numbers */

#define _CRT_SECURE_NO_WARNINGS
#include<stdio.h>
int main()
{
    float this_is_a_number1, this_is_a_number2, total;

    printf("Please enter a number:\n ");
    scanf("%f", &this_is_a_number1); /* read decimal number in */
    printf("You entered %f\n", this_is_a_number1);

    printf("Please enter another number: \n");
    scanf("%f", &this_is_a_number2); /* read decimal number in */
    printf("You entered %f\n", this_is_a_number2);
```

```
        total = this_is_a_number1 + this_is_a_number2;/* add the numbers */
        printf("total is %f\n", total);

        return 0;
}
```

Type in and compile this program. Test the program using decimal numbers (positive and negative). Note that in our scanf and printf, we use %f rather than %d that we used in our previous program. %d means we want to print or read an integer and %f says that we want to read or print a floating point number.

So now we can add two numbers. What about multiplying?

Multiply Two Numbers

```
#define _CRT_SECURE_NO_WARNINGS
#include <stdio.h>

/*  multiply two floating point numbers */

int main()
{
        float this_is_a_number1, this_is_a_number2, total;

        printf("Please enter a number:\n ");
        scanf("%f", &this_is_a_number1); /* read number in */
        printf("You entered %f\n", this_is_a_number1);

        printf("Please enter another number: \n");
        scanf("%f", &this_is_a_number2); /* read number in */
        printf("You entered %f\n", this_is_a_number2);

        total = this_is_a_number1 * this_is_a_number2;/* multiply the numbers */
        printf("product is %f\n", total);

        return 0;
}
```

Type in this code. It is almost identical to the last one except for the multiply sign. Test it with appropriate numbers.

So we can add and multiply. You can probably guess what is coming next. Maybe you could try writing a "divide two numbers" program without looking at the next piece of text.

Divide Two Numbers

Here is the code you will have written.

```
/*  divide two floating point numbers */
#define _CRT_SECURE_NO_WARNINGS
#include <stdio.h>

/*  divide two floating point numbers */

int main()
{
    float this_is_a_number1, this_is_a_number2, total;

    printf("Please enter a number: \n");
    scanf("%f", &this_is_a_number1); /* read number in */
    printf("You entered %f\n", this_is_a_number1);

    printf("Please enter another number:\n ");
    scanf("%f", &this_is_a_number2); /* read number in */
    printf("You entered %f\n", this_is_a_number2);

    total = this_is_a_number1 / this_is_a_number2;/* divide the numbers */
    printf("quotient is %f\n", total);

    return 0;
}
```

If you have not already done it... type in the code, compile it, and test it. As this is a division, you could try as one of your tests dividing by zero.

Forloops

When we were doing our two numbers program, it would have been a bit of a chore to do a similar thing with, say, ten numbers. We could have done it by repeating similar code ten times. We can make this a bit simpler by writing one piece of code but then looping round the same piece of code ten times. This is called a "forloop."

Here is an example of how a forloop can help us.

```
#define _CRT_SECURE_NO_WARNINGS
#include<stdio.h>
/* demonstrate a forloop */
main()

{
        float this_is_a_number, total;
        int i;

        total = 0;

        /* forloop goes round 10 times */
        for (i = 0;i < 10;i++)
        {
                printf("Please enter a number:\n ");
                scanf("%f", &this_is_a_number); /* read number in */
                total = total + this_is_a_number;

        }
        printf("Total Sum is = %f\n", total);
}
```

The format of the for statement is

for(initial value; final value; increment)

The code to go round the loop is contained with the { after the for statement and the } after the statements.

Within the for statement, the variable i is used as the variable to be incremented and tested while going through the loop. Its initial value of i is 0 as shown in the first part of the for statement; then each time the code is completed within the loop, 1 gets added to i (this is what i++ does). After each loop a test is made to see if the i value has reached

10 (this is the i<10 part). When it does, the loop stops. So in this case the code in the loop is executed ten times. Within the code the user is asked to enter a number. This gets added into total in each loop, then the final value is printed out.

We could have achieved the same output if we had written out each of the three lines within the forloop ten times but, as you can see, this saves time and space and is easy to follow in the code. Also, just imagine if you wanted to round the loop 1000 times. All that you would need to change in the code to do that would be i<10 in the for command to i<1000.

Flowcharts

When you are first designing your program, you may find yourself jotting down notes to remind you what to do and when. There are diagrams that can be helpful here. These are called flowcharts. Flowcharts are a useful tool to use here to help with your understanding of the logical sequences of the programs.

There are sets of shapes that are generally used in flowcharts. Some organizations have specific meanings to specific shapes. This is useful if somebody from a different organization looks at the flowchart. Knowing what its shapes mean is therefore useful. Generally the shapes and their usual meanings are shown in Figure 1-1.

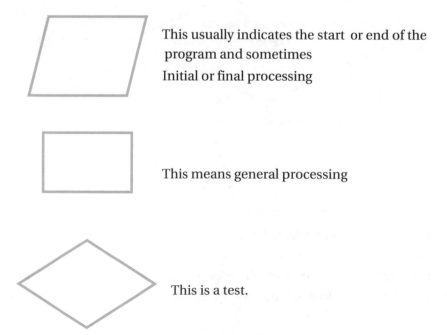

This usually indicates the start or end of the
 program and sometimes
Initial or final processing

This means general processing

This is a test.

Figure 1-1. *Flowchart symbol general meaning*

9

Figure 1-2 shows an example.

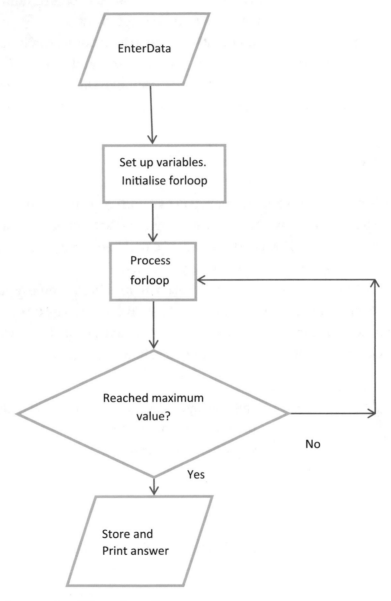

Figure 1-2. *Example of flowchart logic*

This forloop could have been used for your previous program. Start at the top and follow the lines down. The logic of the forloop should be the same as the logic of your program.

You can go even further with loops and have one loop contained inside another. This is called a "nested forloop."

Have a look at this program.

```c
#define _CRT_SECURE_NO_WARNINGS
#include<stdio.h>
/* demonstrate a nested forloop */
main()

{

    float this_is_a_number, total;
    int i, j;

    total = 0;
/* outer forloop goes round 10 times */
    for (i = 0;i < 10;i++)
    {
        /* inner forloop goes round twice */
        for (j = 0;j < 2;j++)
        {
            printf("Please enter a number:\n ");
            scanf("%f", &this_is_a_number); /* read number in */
            total = total + this_is_a_number;
        }

    }
    printf("Total Sum is = %f\n", total);
}
```

This is similar to your previous forloop program except that inside your original forloop, we have another one. So that each time the program enters the outside loop, it does the three lines of code of the inside loop. It goes round the outside loop ten times and the inside loop twice, so in total it does the three lines of code 2*10 = 20 times.

Figure 1-3 shows the flowchart for this program.

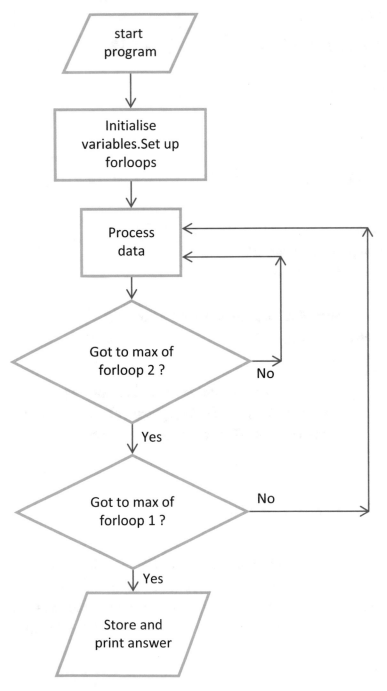

Figure 1-3. *A nested forloop*

Create the program and test it. You may have spotted that you could have achieved the same thing here by just having one forloop with a limit of 20. Quite correct, but in reality we would do other things within the outside loop as well as having our inner loop.

Do Loops

There is another method of doing a similar thing to a forloop, but it is formatted slightly differently. The loop says "do" – then within {}, again, contains a series of commands, ending with "while ..." where the "..." is just a condition to be true. When the condition is not true, it drops out of the loop. So we are using a "do" loop to do the same thing as our first forloop program. The i++ instruction in the do loop just adds 1 to whatever i currently contains. To subtract 1 it's just i--.

```
#define _CRT_SECURE_NO_WARNINGS
#include<stdio.h>
/* demonstrate a do loop */
main()

{
        float this_is_a_number, total;
        int i;

        total = 0;
        i = 0;
        /* do loop goes round until the value of i reaches 10 */
        do {

                printf("Please enter a number:\n ");
                scanf("%f", &this_is_a_number);
                total = total + this_is_a_number;
                i++;

        }while( i < 10);
        printf("Total Sum is = %f\n", total);
}
```

You should find that you get the same result as your forloop program.

Switch Instruction

Another instruction that is useful in C is switch. This takes a value and jumps to an appropriate position in the code depending on the value. In the following program, the user can enter any integer value between 1 and 5.

The switch instruction takes the value, and if it is 1 it jumps to case 1:, if it is 2 it jumps to case 2:, and so on. If the number entered is not an integer from 1 to 5, it drops to the default: case where it outputs an error message.

```c
#define _CRT_SECURE_NO_WARNINGS
#include <stdio.h>
/* Example of a switch operation */
int main()
{
    int this_is_a_number;

    printf("Please enter an integer between 1 and 5:\n ");
    scanf("%d", &this_is_a_number);

    switch (this_is_a_number)
    {

    case 1:
        printf("Case1: Value is: %d", this_is_a_number);
        break;
    case 2:
        printf("Case2: Value is: %d", this_is_a_number);
        break;
    case 3:
        printf("Case3: Value is: %d", this_is_a_number);
        break;
    case 4:
        printf("Case4: Value is: %d", this_is_a_number);
        break;
```

```
case 5:
        printf("Case5: Value is: %d", this_is_a_number);
        break;
default:
        printf("Error Value is: %d", this_is_a_number); /* The number
        entered was not between 1 and 5 */
}
return 0;
}
```

You can do a similar thing but using specific characters rather than numbers. You then jump to the appropriate place using the character as the case name, for example, if you type in a, then you jump to case a:

Type in and test the following program. This code expects the user to type in a lowercase character a, b, c, d, or e. Anything else goes to the default option.

```
#define _CRT_SECURE_NO_WARNINGS
#include <stdio.h>
/* Example of using a switch on characters */
int main()
{

    char this_is_a_character;

    printf("Please enter character a,b,c,d or e:\n ");
    scanf("%c", &this_is_a_character);

    switch (this_is_a_character)
    {
    case 'a':
            printf("a entered");
            break;
    case 'b':
            printf("b entered");
            break;
```

```
        case 'c':
                printf("c entered");
                break;
        case 'd':
                printf("d entered");
                break;
        case 'e':
                printf("e entered");
                break;
        default:
                printf("Default ");
        }
        return 0;
}
```

If Then Else

When a decision has to be made in your program to either do one operation or the other, we use if statements.

These are fairly straightforward. Basically we say

```
if (something is true)
      Perform a task
```

This is the basic form of if.

We can extend this to say

```
if (something is true)
      Perform a task
else
      Perform a different task
```

Here is some C code to demonstrate this.

```c
#include <stdio.h>
/* Example of an if operation */
int main()
{
    int this_is_a_number;

    printf( "Please enter an integer between 1 and 10:\n " );
    scanf( "%d", &this_is_a_number );

    if (this_is_a_number <6)
            printf( "This number is less than 6;\n " );

    printf( "Please enter an integer between 10 and 20:\n " );
    scanf( "%d", &this_is_a_number );

    if (this_is_a_number <16)
            printf( "This number is less than 16\n " );
    else
            printf( "This number is greater than 15\n " );

    return 0;
}
```

Create and test your program. When you are testing, it is good practice to test to each limit and to even enter incorrect data. Here there is no check to see if you really do enter within the ranges specified. You could add a test yourself.

There is an extension of the "if then else" type of command. This is the "if then else if" where you add an extra level of ifs. Following is an extension of your last program to add this.

If Then Else If

```c
#define _CRT_SECURE_NO_WARNINGS
#include <stdio.h>
/* Example of an if then else if operation */
int main()
{
    int this_is_a_number;

    printf("Please enter an integer between 1 and 10:\n ");
    scanf("%d", &this_is_a_number);

    if (this_is_a_number < 6)
            printf("This number is less than 6;\n ");

    printf("Please enter an integer between 10 and 20:\n ");
    scanf("%d", &this_is_a_number);

    if (this_is_a_number < 16)
    {
            printf("This number is less than 16\n ");
    }
    else if (this_is_a_number == 20)
    {
            printf("This number is 20\n ");
    }
    else
    {
            printf("This number is greater than 15\n ");
    }

    return 0;
}
```

The program does the same if as the previous one, but instead of just an else following it, it does else if to test another option. So here it tests if the number entered was less than 16. If it was, it prints "This number is less than 16"; otherwise, it then tests if the number equals 20. If it is, it prints out "This number is 20". Otherwise, it prints out "This number is greater than 15 but not 20".

Data Arrays

There is another way of storing data in our programs rather than in just individual locations. These are called "arrays." They can be defined as "int arr" where all the elements of the array are integers. They can be "char arr" where all the elements are characters. There are also other types which we will see later. We define an integer array with the length of the array which we insert in square brackets, for example, int arr[8] for an array of 8 elements.

The following program shows us how to read in 8 integers and store them in an array.

```
#define _CRT_SECURE_NO_WARNINGS
#include<stdio.h>
/* program to show array use */

int main()

{

    int arr1[8];/* define an array of 8 integers */
    int i;

    printf("enter 8 integer numbers\n");

    for (i = 0;i < 8;i++)
    {
        scanf("%d", &arr1[i]);/* read into arr1[i] */
    }
    printf("Your 8 numbers are \n");

    for (i = 0;i < 8;i++)
    {
        printf("%d ", arr1[i]);
    }
    printf("\n");

}
```

Create this program and test it. It will read the eight characters you enter and store them in the array "arr1". It then reads arr1 and prints out its contents.

To read and write characters into our array, we define it as "char arr" and notice that we use %c in our scanf and printf because %c expects characters and %d expects integers.

```c
#define _CRT_SECURE_NO_WARNINGS
#include<stdio.h>
/* program to show character array use */

int main()

{

    char arr2[10];/* define array of 10 characters */
    int i;

    printf("enter 10 characters \n");

    for (i = 0;i < 10;i++)
    {
        scanf("%c", &arr2[i]);
    }
    printf("Your 10 characters are \n");

    for (i = 0;i < 10;i++)
    {
        printf("%c ", arr2[i]);
    }
    printf("\n");

}
```

Arrays are really useful when we are writing software to solve mathematics problems. We can extend our ideas we have just learned. If we say that our int array we have just used is in one dimension (i.e., numbers in a line), we can have a two-dimensional array (like numbers in a matrix.)

Following is a program that allows you to enter data into a two-dimensional array. It can have a maximum of 8 integers in one part and 7 in the other part. This is defined here as int arr1[7][8]. You can picture it like this.

1	2	3	4	5	6	7	8
4	3	4	5	6	7	8	9
0	4	5	6	7	8	9	10
9	5	6	7	8	9	10	11
3	7	8	9	10	11	12	13
8	8	9	10	11	12	13	14
6	9	10	11	12	13	14	15

This can be referred to as a 7x8 array (like a 7x8 matrix in mathematics). The following code reads data into the array.

```
#define _CRT_SECURE_NO_WARNINGS
#include<stdio.h>

/* example of a 2D array test*/
int main()

{
    int arr1[7][8];/* 2D array */

    int i, j, k, l;

    printf("enter number of rows and columns (max 7 rows max 8
    columns) \n");
    scanf("%d %d", &k, &l);
    if (k > 7 || l > 8)
    {
        printf("error – max of 8 for rows or 7 for columns \n");

    }

    else
    {
        printf("enter array\n");
        for (i = 0;i < k;i++)
        {
```

```
            for (j = 0;j < l;j++)
            {
                    scanf("%d", &arr1[i][j]);
            }
        }
        printf("Your array is \n");
        for (i = 0;i < k;i++)
        {
                for (j = 0;j < l;j++)
                {
                        printf("%d ", arr1[i][j]);
                }
                printf("\n");

        }
    }

}
```

There are a few new ideas in this program. As well as having our two-dimensional array, we also have examples of a nested forloop as seen earlier. We also see something which is a really useful thing to use in your programs. This is called "data vetting." If you look at the definition of our array, its first part has 7 integers and its second has 8 integers. If the user tried to enter more than 8, it would cause the program to fail with an error. We can prevent this by checking that the user does not enter more than the maximum expected number of integers for each part. This is what the first "if" statement does. The first part of the program stores the number of "rows" into k and the number of columns into l. The if statement says that if the number of rows is greater than 7 or the number of columns is greater than 8, then it outputs an error message and terminates the program. The symbol "||" means "or."

The 2-D array stores row by row. So if you enter the data shown earlier and print out the first row, then you should get 1 2 3 4 5 6 7 8. You can write a quick test program to do this.

```
#define _CRT_SECURE_NO_WARNINGS
#include<stdio.h>

/* example of a 2D array test*/
int main()
```

```c
{
    int arr1[7][8];

    int i, j, k, l;

    printf("enter number of rows and columns (max 7 rows max 8
    columns) \n");
    scanf("%d %d", &k, &l);
    if (k > 7 || l > 8)
    {
        printf("error - max of 8 for rows or 7 for columns \n");

    }

    else
    {
        printf("enter array\n");
        for (i = 0;i < k;i++)
        {
            for (j = 0;j < l;j++)
            {
                scanf("%d", &arr1[i][j]);
            }
        }
        printf("Your array is \n");
        for (i = 0;i < k;i++)
        {
            for (j = 0;j < l;j++)
            {
                printf("%d ", arr1[i][j]);
            }
            printf("\n");
        }
    }
    printf("first row of array\n");
```

```
for (j = 0;j < 1;j++)
{
        printf("%d ", arr1[0][j]);
}
printf("\n");
}
```

This is the same as your 2-D array program except that at the end it does an extra bit.

```
for(j=0;j<k;j++)
{
        printf("%d ",arr1[0][j]);
}
```

This just prints out arr[0][0], arr[0][1], arr[0][2], arr[0][3], arr[0][4], arr[0][5], arr[0][6], arr[0][7]. This is how the data is stored in a 2-D array. If you wanted the second row, you just need to change the printf("%d ",arr1[0][j]); in the last forloop to printf("%d ",arr1[1][j]);

Two-dimensional arrays are vital when you write programs to perform operations on matrices later in this book.

Functions

Sometimes when you are writing your programs, you will find that you may end up writing similar lines of code in different places in the program. You can make this easier to do and easier for other people to follow what your code does if you put these similar lines of code in a separate place and just call them when you need them. This separate set of code is called a function. If the function has to do slightly different things each time it gets called, this is fine as you can call the function with a different parameter each time you call it. The following code will demonstrate this. It is a fairly trivial piece of code but it illustrates the point.

```
#define _CRT_SECURE_NO_WARNINGS
#include <stdio.h>

/* This code demonstrates what a function does */
/* The function here compares two numbers and says which is bigger */
/* The user enters three numbers and gets told which is bigger than which !*/
```

```
     void myfunction(int a,int b); /* decaration of your function and its
     parameters */

     int first , second, third;
main()
{
     printf( "Please enter first integer number: " );
     scanf( "%d", &first );
      printf( "Please enter second integer number: " );
     scanf( "%d", &second );
      printf( "Please enter third integer number: " );
     scanf( "%d", &third );

     myfunction(first , second);
     myfunction(first , third);
     myfunction(second , third);
}
void myfunction(int a,int b)
/* the function is outside the main{} part of the program */
/* The function just compares the two parameters, a and b, and says which
is greater*/
{

     if(a>b)
            printf("%d is greater than %d\n", a,b);
     else if (a<b)
            printf("%d is greater than %d\n", b,a);
     else
            printf("%d and %d are equal\n", a,b);

}
```

The function here is called myfunction. Notice that it is defined outside of main{}. It is declared at the start of the program. The function is given two numbers, a and b. It compares the two numbers and says which is bigger. In the main part of the program, the user is prompted to enter three numbers. These are then entered into the calls to myfunction in the main part of the code.

This is a fairly simple piece of code but it shows how a function can be used.

The following piece of code also shows how functions are used. This code is based on the program you wrote in the 2-D arrays section of this chapter. It prints out specific rows of your 2-D array. One call to the function asks the function to print out the second row of the array, and the other call asks it to print out the first row.

```c
#define _CRT_SECURE_NO_WARNINGS
#include<stdio.h>

/* example of a function*/
void printarow(int row, int cols, int arr[8][8]);
int main()

{
    int arr1[8][8];

    int i, j, rows, cols;

    printf("enter number of rows and columns (max 8 rows max 8 columns)
\n");
    scanf("%d %d", &rows, &cols);
    if (rows > 8 || cols > 8)
    {
        printf("error - max of 8 for rows or columns\n");

    }

    else
    {
        printf("enter array\n");
        for (i = 0;i < rows;i++)
        {
            for (j = 0;j < cols;j++)
            {
                scanf("%d", &arr1[i][j]);
            }
        }
        printf("Your array is \n");
        for (i = 0;i < rows;i++)
        {
```

```
            for (j = 0;j < cols;j++)
            {
                  printf("%d ", arr1[i][j]);
            }
            printf("\n");

        }
    }
    printarow(2, cols, arr1);/* This calls to print out row 2
    only(assumes that you have at least 2 rows) */
    printf("\n");
    printarow(1, cols, arr1);/* This calls to print out row 1 only */
    printf("\n");
}

void printarow(int row, int cols, int arr[8][8])

/* this is a function which can be called from anywhere in the program */
/* and can be called as often as you want to */
/* If you need to do the same type of thing many times it saves you */
/* having to write out the same code repeatedly. All you need to */
/* is call the function */

{
    int j;
    printf("row %d is ", row);
    for (j = 0;j < cols;j++)
    {
          printf("%d ", arr[row - 1][j]);
    }

}
```

Notice that the array name used in the function does not have to be the same as that used in main{}. In the instruction if(rows>7 || cols>8), the || means OR. So here we are saying if the user has specified more than seven rows or more than eight columns, then we print an error and stop the program. At the end of the chapter, the common arithmetic and logical symbols used in C are listed.

Create and test this program. The code assumes you have at least two rows. You could amend the code to call `printarow` as many times as you want to.

A function can return a value to the caller. The following code demonstrates this.

```c
/* Function which returns an answer  */
/* finds the pupil in one year of the school with the highest marks */

#include <stdio.h>
double getmarks(double pupils[]);

int main()
{
    double pupil;
    /* Array with marks for class is preset in the main part of the
    program */
    double marks[] = { 10.6, 23.7, 67.9, 93.0, 64.2, 33.8 ,57.5 ,82.2
    ,50.7 ,45.7 };
    /* Call function getmarks. The function returns the max marks which
    is then stored in pupil */
    pupil = getmarks(marks);
    printf("Max mark is  = %f", pupil);
    return 0;

}

double getmarks(double pupils[])
{
    int i;
    double highest;
    highest = 0;
    /* Go through all the pupils in turn and store the highest mark */
    for (i = 0; i < 6; ++i)
    {
            if (highest < pupils[i])
                    highest = pupils[i];

    }
    return highest; /* returns the value in highest to where the function
    was called */

}
```

The function is called getmarks. It returns a value to the point where it was called. In real-life programs, the function will be called many times from different points in the program. This technique is efficient and makes the program easier to follow.

Strings

Strings in C are just like character arrays we have already looked at. The main difference is that the string has a NULL character at the end. This is just to show where the string ends as we have to do things like compare two strings or find the length of the string. To find the length, we have a function written for us in the string.h library, and this needs to have a NULL character at the end. As a result of this, if we are defining a preset string as a character array of a certain length, we need to account for the NULL at the end. So if our string had "Borrow" in it, the word has six characters so our string array would have to have seven characters in its definition to account for the NULL character at the end. When we print a string using printf, we use %s to show it is a string (where we used %d to print an integer or %f to print a floating point number).

Here is a program to check the length of strings (strlen), copy onto another (strcpy), concatenate two strings (strcat), and compare the contents of two strings (strcmp).

Concatenation of two strings is just tagging one string onto the end of the other.

```
#define _CRT_SECURE_NO_WARNINGS
#include <stdio.h>
#include <string.h>
/* Program to demonstrate string operations strlen, strcpy, strcat, strcmp */

int main() {
    char borrow[7] = { 'b', 'o', 'r', 'r', 'o', 'w','\0' };
    char string1[32] = "This is string1";
    char string2[16] = "This is string2";
    char string3[16];
    int  len;
    /* Print out the lengths of the strings */
```

```
        len = strlen(string1);
        printf("strlen(string1) :  %d\n", len);
        len = strlen(string2);
        printf("strlen(string2) :  %d\n", len);
        len = strlen(string3);
        printf("strlen(string3) :  %d\n", len);

        /* copy string1 into string3 */

        strcpy(string3, string1);
        printf("strcpy( string3, string1) :  %s\n", string3);
        len = strlen(string3);
        printf("strlen(string3) after copy of string1 into string3 :  %d\n", len);
        /* Compare string1 and string3 (these should be the same)*/
        if (strcmp(string1, string3) == 0)
                printf("strings are the same\n");

        /* concatenates string1 and string2 */
        strcat(string1, string2);
        printf("strcat( string1, string2):  %s\n", string1);
        /* total length of string1 after concatenation */
        len = strlen(string1);
        printf("strlen(string1) after cat of string2 onto string1 :  %d\n", len);
        printf("String as predefined quoted chars: %s\n", borrow);

        return 0;
}
```

In strlen the function returns the length of the string.

In strcpy the function copies the second string in the command to the first.

In strcmp the function compares the contents of the two strings and returns 0 if they are equal.

In strcat the function tags the second string onto the end of the first string.

Mathematical Functions

There are many mathematical functions that you can access in C. The ones featured here are only a sample. A more comprehensive list can be found in the appendix. The functions featured here are available in most scientific calculators. However, I have selected these to illustrate how you can access these functions as part of much more complicated calculations that you will have in your programs.

The following set are the common trigonometric functions cosine, sine, tangent, and their inverses. We include the `math.h` library which contains these functions.

```c
#define _CRT_SECURE_NO_WARNINGS
#include <stdio.h>
#include <math.h>
/* Illustration of the common trigonometric functions */

int main()
{
#define PI 3.14159265
    double angle, radianno, answer;

    /* The cosine function */
    printf("cosine function:\n ");
    printf("Please enter angle in degrees:\n ");
    scanf("%lf", &angle);
    printf("You entered %lf\n", angle);
    radianno = angle * (2 * PI / 360);
    answer = cos(radianno);
    printf("cos of %lf is %lf\n", angle, answer);

    /* The sine function */
    printf("sine function:\n ");
    printf("Please enter angle in degrees:\n ");
    scanf("%lf", &angle);
    printf("You entered %lf\n", angle);
    radianno = angle * (2 * PI / 360);
    answer = sin(radianno);
    printf("sin of %lf is %lf\n", angle, answer);
```

```
    /* The tangent function */
    printf("tangent function:\n ");
    printf("Please enter angle in degrees:\n ");
    scanf("%lf", &angle);
    printf("You entered %lf\n", angle);
    radianno = angle * (2 * PI / 360);
    answer = tan(radianno);
    printf("tan of %lf is %lf\n", angle, answer);

    return 0;

}
```

The next program finds the arccos, arcsin, and arctan. So here the user has to enter the cosine of an angle and the program works out the angle, and similarly for arcsin and arctan. The functions the program uses are arccos, arcsin, and arctan. Again, the angles that the function returns will be radians. You can display this value to the user if you wish, but make sure that you print "radians" in your printf function. Here we convert the angle into degrees and tell the user this.

```
#define _CRT_SECURE_NO_WARNINGS
#include <stdio.h>
#include <math.h>

int main()
{
#define PI 3.14159265
    double radianno, answer, arccos, arcsin, arctan;

    /* The arccos function */
    printf("arccos function:\n ");
    printf("Please enter arccos:\n ");
    scanf("%lf", &arccos);
    printf("You entered %lf\n", arccos);
    radianno = acos(arccos);
    answer = radianno * (360 / (2 * PI));
    printf("arccos of %lf in degrees is %lf\n", arccos, answer);
```

```
/* The arcsin function */
printf("arcsin function:\n ");
printf("Please enter arcsin:\n ");
scanf("%lf", &arcsin);
printf("You entered %lf\n", arcsin);
radianno = asin(arcsin);

answer = radianno * (360 / (2 * PI));
printf("arcsin of %lf in degrees is %lf\n", arcsin, answer);

/* The arctan function */
printf("arctan function:\n ");
printf("Please enter arctan:\n ");
scanf("%lf", &arctan);
printf("You entered %lf\n", arctan);
radianno = atan(arctan);
answer = radianno * (360 / (2 * PI));
printf("arctan of %lf in degrees is %lf\n", arctan, answer);

return 0;

}
```

We use #define PI 3.14159265. This is similar to a variable definition except that we preset its value, and this value is used as a constant in the program.

The code here asks the user to enter an angle in degrees. The functions to find the cosine, sine, and tangent of an angle are called cos, sin, and tan. The functions expect the angles to be in radians. Here the user can enter the angle in degrees and the program converts this value to radians.

The next set of functions is to find the exponent of a number (exp to the power of whatever the number is), to find the natural logarithm of a number (ln of number), and to find the log to base 10 of a number (log of the number).

In C the natural logarithm function is log and the log to base 10 is log10.

```
#define _CRT_SECURE_NO_WARNINGS
#include <stdio.h>
#include <math.h>
/* showing use of exp, log and log10 functions */
```

```
int main()
{
    double answer, expno, natlog, lb10;

    /* find exponent of entered number */
    printf("exponental function:\n ");
    printf("Please enter number:\n ");
    scanf("%lf", &expno);
    printf("You entered %lf\n", expno);

    answer = exp(expno);
    printf("exponent of %lf is %lf\n", expno, answer);

    /* find natural logarithm of entered number */
    printf("natural logarithm function:\n ");
    printf("Please enter number:\n ");
    scanf("%lf", &natlog);
    printf("You entered %lf\n", natlog);
    answer = log(natlog);
    printf("natural logarithm of %lf is %lf\n", natlog, answer);

    /* find log to base 10 of entered number */
    printf("log to base 10 function:\n ");
    printf("Please enter number:\n ");
    scanf("%lf", &lb10);
    printf("You entered %lf\n", lb10);
    answer = log10(lb10);

    printf("log to base 10 of %lf is %lf\n", lb10, answer);
}
```

The final program here containing mathematical functions finds the power of a number (you enter the number and the power you want it raising to), the square root of a number, and the absolute value of a number. When using the power function, pow, you can specify inverse powers. So x^2 would be pow(x,2) and x^5 would be pow(x,5). Similarly $1/x^2$ (which can also be written as x^{-2}) would be pow(x,-2). When your function includes coefficients (the number before the x term), sometimes it is necessary to simplify. For example, if you had $3/4x^3$, it is best to rewrite this as $0.75/x^3$ so you won't have to worry about specifying separate numbers (here the 3 and the 4).

```c
#define _CRT_SECURE_NO_WARNINGS
#include <stdio.h>
#include <math.h>
/* showing use of pow, sqrt and fabs functions */
int main()
{
    double answer, pownum, power, sqroot, fabsno;

    /* find x raised to power y number */
    printf("power:\n ");
    printf("Please enter number:\n ");
    scanf("%lf", &pownum);
    printf("You entered %lf\n", pownum);
    printf("Please enter power:\n ");
    scanf("%lf", &power);
    printf("You entered %lf\n", power);

    answer = pow(pownum, power);
    printf("%lf raised to power %lf is %lf\n", pownum, power, answer);

    /* find square root of number */

    printf("square root:\n ");
    printf("Please enter number:\n ");
    scanf("%lf", &sqroot);
    printf("You entered %lf\n", sqroot);

    answer = sqrt(sqroot);
    printf("The square root of %lf is %lf\n", sqroot, answer);

    /* find absolute value of number */
    printf("absolute value:\n ");
    printf("Please enter number:\n ");
    scanf("%lf", &fabsno);
    printf("You entered %lf\n", fabsno);

    answer = fabs(fabsno);
    printf("The absolute value of %lf is %lf\n", fabsno, answer);

}
```

This program is fairly straightforward. In each of the four programs, when you are testing them, enter data that you know the correct answers to. Again, when testing it is vital (when you are writing code for real) to test the software to the limit, that is, find values that may be out of the range you intended and enter those. You have to act as a "devil's advocate" and, basically, try to cause the program to fail. When your software is used in a real-life situation for any long period of time, this kind of thing happens anyway so you have to either anticipate it happening and put in code to avoid it from happening (data vetting) or warn the user.

Structures

The variables used up to now have just been singly named variables of a certain type. Another type of variable is a structure. This is a variable that contains separate variables within it. If you imagine a file containing details of a student at a college, the details of each student might be their name, their student ID, and possibly their last examination mark. So, in a paper file, these may be held like this:

id
Name
Percent

So there would be an entry like this in the file for each student.

Here is a program which declares such a structure. It then assigns variable names s1 and s2 to have that type of definition. Then it gives each structure values, then prints them out.

```c
/* Structure example program */
#define _CRT_SECURE_NO_WARNINGS
#include<stdio.h>
#include<string.h>

/* define the structure */
struct Student {
    int id;
    char name[16];
    double percent;
};
```

```c
int main() {
      /* define two data locations of type "student" */
      struct Student s1, s2;

      /* Assign values to the s1 structure */

      s1.id = 56;
      strcpy(s1.name, "Rob Smith");
      s1.percent = 67.400000;
      /* Print out structure s1 */

      printf("\nid : %d", s1.id);
      printf("\nName : %s", s1.name);
      printf("\nPercent : %lf", s1.percent);

      /* Assign values to the s2 structure */

      s2.id = 73;
      strcpy(s2.name, "Mary Gallagher");
      s2.percent = 93.800000;

      /* Print out structure s1 */

      printf("\nid : %d", s2.id);
      printf("\nName : %s", s2.name);
      printf("\nPercent : %lf", s2.percent);

      return (0);
}
```

This can be extended so instead of defining individual entries (s1 and s2), we can define a larger number in one definition. In the following example, we define five items in the array year9. Then we refer to the first student entry as year9[0], the second student entry as year9[1], and so on. Look at the following code.

```c
/* Structure example program (extended structure)*/
#define _CRT_SECURE_NO_WARNINGS
#include<stdio.h>
/* define the structure */
struct Student {
```

```c
    int id;
    char name[16];
    double percent;
};
int main() {
        int i;
    /* define 5 data locations of type "student" */
        struct Student year9[5];
        for(i=0; i<5; i++)
        {
                /* Assign values to the structure */
                printf("enter student ID\n");
                scanf("%d",&year9[i].id);
                printf("enter student name\n");
                scanf("%s",year9[i].name);
                printf("enter student percent\n");
                scanf("%lf",&year9[i].percent);
        }
        for(i=0; i<5; i++)
        {
                /* Print out structure s1 */

                printf("\nid : %d", year9[i].id);
                printf("\nName : %s", year9[i].name);
                printf("\nPercent : %lf", year9[i].percent);

        }
    return (0);
}
```

This type of structure definition is vital when you set up files and write them or read them. You will see more of structures in the chapter dealing with file usage.

Size of Variables

There is a useful function in C which tells you the size in bytes of variables on your machine. Sometimes, different compilers or software development tools have different sizes for different structures. The function is called `sizeof`. You supply it with the variable type you want to know the size of, and it returns the answer as the number of bytes.

You can also supply a structure as the parameter if you don't know its size.

A program to do the basics is shown as follows.

```
/* Program to illustrate the use of the sizeof command */

#include <stdio.h >
#include < limits.h >
#include < math.h >

    int main() {

    int sizeofint;
    unsigned int sizeofunsint;
    float sizeoffloat;
    double sizeofdouble;

    printf("storage size for int : %d \n", sizeof(sizeofint));
    printf("storage size for uns int : %d \n", sizeof(sizeofunsint));
    printf("storage size for float : %d \n", sizeof(sizeoffloat));
    printf("storage size for double float: %d \n", sizeof(sizeofdouble));

    return(0);

}
```

This prints out the sizes of an int, an unsigned int, a floating point, and a double floating point.

Goto Command

Under some circumstances, you may want to jump out of your normal sequence of code, for instance, if you discover an error in a sequence of code. In this case, you can define a label and jump to the label from within your sequence of code.

goto is not used frequently in programming (it's actively discouraged in fact). It can quite quickly lead to tangled and unmaintainable code. However, it can be used if you want a quick exit from your program.

This is shown in the following code.

```c
#include <stdio.h> /* Demonstrate a goto statement */
int main()
{
    int i, testvalue;

    testvalue = 2;

    for (i = 0; i < 10; i++)

    {
        if (testvalue == 2)
            goto error;
    }
    printf("Normal Exit from forloop\n");

error:
    printf("testvalue is %d\n", testvalue);
}
```

So here the code would jump to error and output testvalue is 2.

Common Mathematical and Logical Symbols

Here is a quick reference:

= assign

== equals

!= not equal to

< less than

> greater than

<= less than or equal to

>= greater than or equal to

&& logical AND

|| logical OR

! logical NOT

EXERCISES

1. Amend your add two numbers program to read in and add five numbers. Test with the following sets of numbers:

    (i) 2, 15, 213, 51, 8

    (ii) 1234, 2345, 1517, 2, 5205

    (iii) −1234, −2345, −1517, −2, −5205

    (iv) 11994, −10000, −900, −90, −4

2. Amend your add two decimal numbers program to subtract two decimal numbers. Test with the following sets of numbers:

    (i) 12.6, 11.3

    (ii) 11994, 11993.

3. Amend your multiply two numbers program to multiply three numbers. Test with the following sets of numbers:

    (i) 8.0, 3.2, 7.6.

    (ii) 8.0, 2.5, −0.6

    (iii) 66975, 285, −0.0087

    (iv) 395434, 454, 0.00003

    (v) 395454, −871, 0

4. In your divide two numbers program, change it to read in two numbers. Add them. Then read in another number and divide your added total by your third number. Test with these sets of data:

    (i) (10+ 35) / .15

    (ii) (300030 + 4600) / 1486

    (iii) (1610 + 2004) / 2365

5. Now change your program in question 4 to read four numbers. Add the first two together, then add the second two together. Divide your first sum by your second sum. Test with this data:

    (1610 + 2004) / (2005 + 360)

6. For your forloop example, in your forloop program read in an integer number and use it in your for instruction as the limit for the loop. Test your amendment by giving it the same number as your original forloop program.

7. In your nested forloop program, add another nested forloop within the two you already have. Go round your new loop three times. Test your amendment.

8. For your do loop program, change the while statement to limit your do loop for i>10. (Make sure you get your initial value of i correct. What do you think will happen if your initial value is 10?

9. Write a program to extend the data array program so that you enter and store two separate arrays. Then print out the first line of the first array and the first line of the second array.

10. Write a program similar to the one where you return a value from a function. In this case, the function is called to find the average of a set of marks. Set up an array in the main part of the program and initialize it with nine values. Call your function to calculate the average of these numbers. Your function should return the average, and this value is stored in the main part of the program and then printed out.

11. Using the mathematical functions in this chapter, write a program to find the length of AC in the diagram:

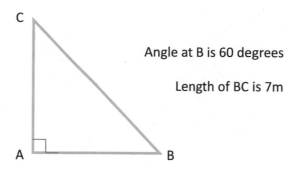

Angle at B is 60 degrees

Length of BC is 7m

12. Similarly to the last question, find the length of YZ in the following diagram.

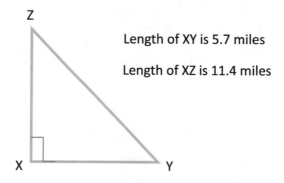

Length of XY is 5.7 miles

Length of XZ is 11.4 miles

13. As in the last question, find the length of MN in the following diagram.

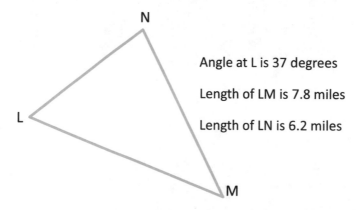

Angle at L is 37 degrees

Length of LM is 7.8 miles

Length of LN is 6.2 miles

14. In the "extended structure" program in the "Structures" section of this chapter, amend the program so that instead of entering data for the two individual cases, you set up a forloop with five iterations. For the first iteration, enter data for student 1. For the second iteration, enter data for student 2 and so on. Then set up another forloop to print all of the data for the five students.

CHAPTER 2

Solving Equations

The first mathematical application which we will write a C program for is solving quadratic equations. We will then write some programs to solve equations of higher powers.

Quadratic Equations

Quadratic equations are equations which have a squared variable as their highest power.

So $x^2 + 2x + 1 = 0$, $2x^2 - 6x + 7 = 0$, and $-x^2 + 2x - 15 = 0$ are all quadratic equations, but $x^3 + 2x^2 - 3x + 7 = 0$, $2x^5 - 3x = 0$, and $3x - 2 = 0$ are not quadratic equations.

The normal analytical methods for solving quadratics are

- Factorizing

- Completing the square

- Using the quadratic formula

Factorizing

Factors of any ordinary whole numbers are just those numbers that will divide into that whole number. So the factors of 15 are 1, 3, 5, and 15. If we are asked to find the factors of a quadratic equation like $x^2 + 5x + 6$, we want to find two algebraic terms that multiply together to give $x^2 + 5x + 6$. In this case the two terms are $(x + 2)$ and $(x + 3)$. You can test this out by multiplying the two terms together.

Now if we were asked to solve the equation $x^2 + 5x + 6 = 0$, we can rewrite the equation as $(x + 2)(x + 3) = 0$. Now if two numbers multiply to give 0, then either one or both of the numbers must be 0. So we can say that either $(x + 2) = 0$ or $(x + 3) = 0$. So rearranging each of these gives $x = -2$ or $x = -3$. These are the factorized solutions of $x^2 + 5x + 6 = 0$.

The quadratic formula can look a bit daunting when you first see it, but it is just derived from the "completing the square" method. This is shown next.

© Philip Joyce 2019
P. Joyce, *Numerical C*, https://doi.org/10.1007/978-1-4842-5064-8_2

Completing the Square

If we have a quadratic equation $x^2 + 8x + 7 = 0$, we can solve it by taking the x^2 term and the $8x$ term and making a square from it. We do this by starting with $(x + ?)^2$. We want to know that we can put where the question mark is so that we can end up with the first part of our equation, $x^2 + 8x$. If we halve the 8 and insert that, we then get $(x + 4)^2$ as our square. Now if we expand this, we get $x^2 + 8x + 16$. But this is 16 more than we want. So we just subtract it. So now we can say $x^2 + 8x$ is the same as $(x + 4)^2 - 16$.

You might think why would you want to do that.

1. Well we can now rewrite our original equation as $(x + 4)^2 - 16 + 7 = 0$ by just replacing $x^2 + 8x$ in our original equation by $(x + 4)^2 - 16$.

2. So now we have $(x + 4)^2 - 16 + 7 = 0$.

3. Rearranging we get $(x + 4)^2 = 16 - 7$.

4. Or $(x + 4)^2 = 9$.

5. So now we square root both sides to get $x = -4 + 3$ or -3.

6. And finally $x = -4 + 3$ or $x = -4 - 3$.

7. Or $x = -1$ or -7.

The advantage in the "completing the square" method is that it is useful when the original equation is difficult to factorize (although in this case we could have factorized it).

All you need to remember is halve your number before the x term and put it in the square.

1. So for $x^2 + 94x - 53 = 0$.

2. We can just write $(x + 47)^2 - 47^2 - 53 = 0$.

3. So $(x + 47)^2 = 53 + 47^2$.

4. So taking square roots $(x + 47) = +-(53 + 47^2)^{1/2}$.

5. So $x = -47 + (53 + 47^2)^{1/2}$.

6. Or $-47 - (53 + 47^2)^{1/2}$.

It's easier if we have the coefficient of x^2 to be 1. So if we had to factorize $2x^2 + 4x - 23 = 0$, it's best to divide both sides by the coefficient of x. Here this is 2, so doing this we get

$x^2 + 2x - 23/2 = 0$

then we can proceed as before.

Quadratic Formula

In the general case when we can have any numbers for the coefficients of x^2 and x and as our constant, we can write

$ax^2 + bx + c = 0$

where a, b, and c are any numbers (although a cannot be 0 as in that case it would not be a quadratic equation).

If we do our normal method for completing the square on this equation

1. First we divide by the number before the x^2 to get
    $x^2 + (b/a)x + (c/a) = 0$.

2. Now we make our square by halving the number before the
    x $(x + (b/2a))^2 - (b/2a)^2$ (not forgetting to take off the square
    of the number).

3. Then we can write $(x + (b/2a))^2 - (b/2a)^2 = -(c/a)$.

4. So $(x + (b/2a))^2 = (b/2a)^2 - (c/a)$.

5. Take the square root of both sides $(x + (b/2a)) = +-((b/2a)^2 - (c/a))^{1/2}$.

6. So $x = -(b/2a) +-((b/2a)^2 - (c/a))^{1/2}$.

7. We can add the fractional terms inside the square root as $b^2/4a^2 - c/a = (b^2 - 4ac)/4a^2$.

8. So we get for the square root term $((b^2 - 4ac)/4a^2)^{1/2}$.

9. We can take the square root of the denominator to be 2a.

10. So now we have $x = -b/2a +- (b^2 - 4ac)^{1/2} /2a$.

11. So collecting the two terms we get $x = (-b +- (b^2 - 4ac)^{1/2})/2a$

which is the quadratic formula.

In our program we only need to ask the user to type in their values of a, b, and c, then we can use the quadratic formula to find our values of x which will be the solution to the quadratic equation.

The following code does this.

```c
/*quad3 - first attempt at quadratic solver*/
#define _CRT_SECURE_NO_WARNINGS
#include <stdio.h>
#include <math.h>
main()
{
    double a, b, c, xa, xb;

    /* prompt and read in coefficients of x^2,x and constant */
    printf("enter a value");
    scanf("%lf", &a);
    printf("enter b value");
    scanf("%lf", &b);
    printf("enter c value");
    scanf("%lf", &c);
    if (pow(b, 2) < 4 * a*c) /* test for real root */
    {
        /* not real root */
        printf("Not a real root");
    }
    else
    {
        /* real root */
        xa = (-b + sqrt(pow(b, 2) - (4 * a*c))) / (2 * a);
        xb = (-b - sqrt(pow(b, 2) - (4 * a*c))) / (2 * a);
        printf("Roots are %lf and %lf", xa, xb);
    }

}
```

Try this code to solve our earlier equation $x^2+8x+7=0$. So enter a=1, b=8, and c=7. Here, pow(b,2) is the mathematical function that raises b to the power of 2. Notice the test for $b^2 < 4ac$.

If this is true, then when we do $b^2 - 4ac$ in the formula, we would get a negative number. So when we try to take the square root of this negative number, the program would fail.

If you are aware of the existence of complex numbers, then you can accommodate the square root of a negative number. The following code shows this.

```c
/*quad1 -  quadratic solver with complex numbers*/
#define _CRT_SECURE_NO_WARNINGS
#include <stdio.h>
#include <math.h>
main()
{

    double a, b, c, xra, xia, xrb, xib, xa, xb;

    /* prompt and read in coefficients of x^2,x and constant */
    printf("enter A value");
    scanf("%lf", &a);
    printf("enter B value");
    scanf("%lf", &b);
    printf("enter C value");
    scanf("%lf", &c);
    if (pow(b, 2) < 4 * a*c) /* test for complex root */
    {
        /* complex root */
        /* switch b^2 and 4ac to find the positive root then add i to
        the answer*/
        printf("complex root\n");
        xra = -b / (2 * a);
        xia = sqrt((4 * a*c) - (pow(b, 2))) / (2 * a);
        xrb = -b / (2 * a);
        xib = -sqrt((4 * a*c) - (pow(b, 2))) / (2 * a);
        printf("Roots are %lf +%lfi and %lf - %lfi", xra, xia, xra, xia);
    }
```

```
    else
    {
            /* real root */
            xa = (-b + sqrt(pow(b, 2) - (4 * a*c))) / (2 * a);
            xb = (-b - sqrt(pow(b, 2) - (4 * a*c))) / (2 * a);
            printf("Roots are %lf and %lf", xa, xb);
    }
}
```

Try entering a=5, b=2, c=1.

You should get the answer x=-0.2+0.4i and x=-0.2-0.4i.

Equations of Higher Powers

Let's look at some equations of higher powers.

Trial and Improvement

In the last section, we looked at ways of solving quadratic equations. This is relatively simple. Things get more complicated with higher powers of x. There is a technique which looks a bit non-analytical at first sight. In some ways it is non-analytical, but this is where the power of computers takes over. You can use the Trial and Improvement technique with pen and paper.

What you do is guess a value of x and put it into the equation. If the answer we get is higher than the one we want, we try a lower value of x. If this gives a lower answer than the first, then we know that the actual correct value of x must be in between the two values we tried. If you try this method, the best way is to use a table to clarify what you are doing.

If we want to solve the equation $x^3 + x - 12 = 0$, firstly we need to know what the graph looks like. Figure 2-1 is a screenshot of the curve from the Graph package.

The curve is for $y = x^3 + x - 12$, so if our equation is $x^3 + x - 12 = 0$, then we are finding x when y = 0. So look at where the curve crosses the x-axis.

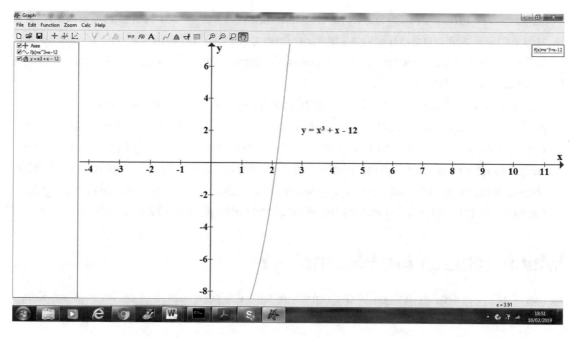

Figure 2-1. *Screenshot of y = x³ + x – 12*

This shows that our solution must be between x=2 and x = 3.

We try substituting x = 2 into x³ + x –12.

We get the answer –2 which looks about right from the graph. Then we try x = 3 and we get +18. So we can now try values in between x = 2 and x = 3 to see if these are above or below 0.

We can draw a table to clarify this (Figure 2-2).

Value of x	x³+x–12	Comment
2	–2	Too low
3	18	Too high
2.5	6.125	Too high
2.2	0.848	Too high
2.1	–0.639	Too low

Figure 2-2. *Table of Trial and Improvement steps*

You can see the value of x must be between 2.1 and 2.2. If we carry on by entering a value of x between our too high and too low limits, we will be getting closer to our target of 0. If the answer turns out to be 25 places of decimals, then this could take some time. A C program can do it in seconds.

It may have crossed your mind that this technique of going round and round inserting different values and testing them sounds a bit like what we have done using our forloops earlier. This is exactly what we use here. We can specify how many times we want to go round our forloop testing our "too high" and "too low conditions." What we need to know in advance is roughly where our solution is. We have this already for $x^3 + x - 12 = 0$, but in the next example, we want to solve $x^2 - 8x + 12 = 0$.

Which Solution Are We Finding?

In our first program, we will code the equation into the program, but on later examples, we will be able to enter a different equation each time. In this program, the equation to be solved is $x^2 - 8x - 13 = 0$.

Figure 2-3 is a screenshot of this curve.

We can see that there are two solutions. We will look at the one to the left of the curve. The solution is between x = -1 and x = -2. By looking at the curve, a value of x=-1 will give us a fairly low position on the curve, whereas x=-2 is high above the x-axis. So our starting low value will be x=-1 and our starting high value will be x=-2. Note that the "low" value is not necessarily the smaller of your two values of x. Here, for the solution to the left of the curve, x=-2 is the high value and x=-1 is the low value because x=-2 is the value of x where the curve is above the x-axis and x=-1 is where the curve is below the x-axis.

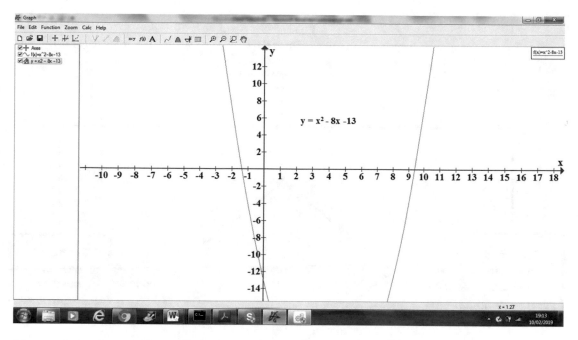

Figure 2-3. *Screenshot of y = x² –8x –13*

Figure 2-4 is a flowchart showing the basic logic of this procedure.

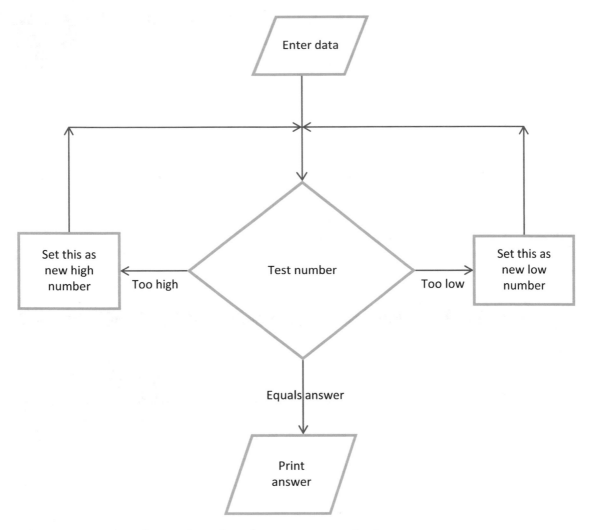

Figure 2-4. *Flowchart of Trial and Improvement logic*

The code is shown as follows.

```
/*      trialimp */
/*      program uses x^2 -8x - 13 =0*/
/*      equation is solved by trial & improvement */
#define _CRT_SECURE_NO_WARNINGS
#include <stdio.h>
#include <math.h>
main()
```

```
{

    float p2, p1, lower, upper;

    double testhigh, testlow, testvalue, middle;
    float constval;
    int i, iterations;

    /* Your curve should cross the x-axis */
    /* Here, your lower value is a value of x where your curve is below
    the x-axis */
    /* Your upper value is a value of x where your curve is above the
    x-axis */
    /* Both values should be close to where your curve crosses the x-axis */

    printf("enter initial lower value");
    scanf("%f", &lower);
    printf("enter initial upper value");
    scanf("%f", &upper);
    printf("enter number of iterations");
    scanf("%d", &iterations);

    /* Preset constant values   */
    p2 = 1;/* coefficient of power of x squared */
    p1 = -8;/* coefficient of x */
    constval = -13; /* numeric constant*/

    testlow = lower;
    testhigh = upper;
    printf("Equation is:-%f x**2 %f x %f=0\n", p2, p1, constval);
    /*printf("%f x**2 %f x= %f\n", p2,p1,constval);*/
    for (i = 0;i < iterations;i++)
    {
            middle = (testhigh + testlow) / 2;
            testvalue = pow(middle, 2) - 8 * middle - 13;
```

```
        if (testvalue == 0)
        {
                printf("x is %f", middle);
                return(0);
        }
        if (testvalue > 0)
        {
                testhigh = middle; /* replace upper value with this one */
        }
        else
        {
                testlow = middle; /* replace lower value with this one */

        }
    }
    printf("x is %f", middle);

}
```

The second solution is between 9 and 10. Looking at the graph, 9 would give us a value below 0 and 10 would give us a value above 0. So here 9 is our lower value and 10 is our upper value.

Three Solutions

Our next example is a cubic. Its equation is $y = x^3 + 2x^2 - x$. The screenshot for this curve is shown in Figure 2-5.

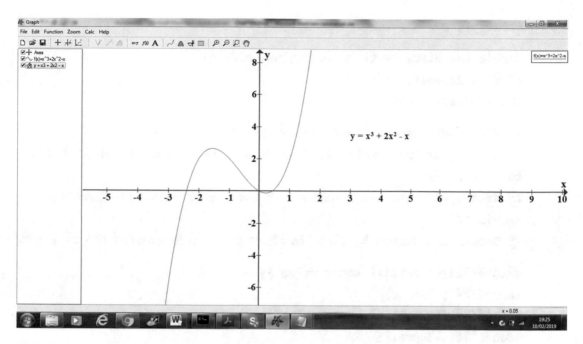

Figure 2-5. *Screenshot of* $y = x^3 + 2x^2 - x$

The curve crosses the x-axis at three places. The one to the left is between x = –2 and x = –3. The other two are both between x = 0 and x = 1. We have to zoom in a little closer to find our upper and lower bounds here. We find that they are –0.1 and 0.1 for the left solution and 0.3 and 0.5 for the right solution. This highlights a potential problem with the Trial and Improvement method. You cannot always take any upper and lower value. You need to have an idea of the shape of the graph and where it crosses the x-axis; otherwise, you will get incorrect answers.

The code for this function is shown as follows.

```
/*      trialimp */
/*      program uses x^3 +2x^2 - x =0*/
/*      equation is solved by trial & improvement */
#define _CRT_SECURE_NO_WARNINGS
#include <stdio.h>
#include <math.h>
main()
{
```

```
float p3, p2, p1, lower, upper;

double testhigh, testlow, testvalue, middle;
float constval;
int i, iterations;

/* Your curve should cross the x-axis */
/* Here, your lower value is a value of x where your curve is below
the x-axis */
/* Your upper value is a value of x where your curve is above the
x-axis */
/* Both values should be close to where your curve crosses the x-axis */

printf("enter initial lower value");
scanf("%f", &lower);
printf("enter initial upper value");
scanf("%f", &upper);
printf("enter number of iterations");
scanf("%d", &iterations);

/* Preset constant values  */
p3 = 1;/* coefficient of power of x power 3 */
p2 = 2;/* coefficient of power of x squared */
p1 = -1;/* coefficient of x */
constval = 0; /* numeric constant*/

testlow = lower;
testhigh = upper;
/*printf("Equation is:-%f x**2 %f x %f=0\n", p3,p2,constval);*/

printf("Equation is:%f x**3 %f x**2 %f x= 0\n", p3, p2, p1);
for (i = 0;i < iterations;i++)
{
        middle = (testhigh + testlow) / 2;
        testvalue = pow(middle, 3) + 2 * pow(middle, 2) - middle;
```

```
        if (testvalue == 0)
        {
                printf("x is %f", middle);
                /*exit(1);*/
                return(0);/*  test */
        }
        if (testvalue > 0)
        {
                testhigh = middle;/* replace upper value with this one */

        }
        else
        {
                testlow = middle;/* replace lower value with this one */

        }
    }
    printf("x is %f", middle);
}
```

As you can see, the code is almost identical to the previous one. It just has a different function specified.

User-Entered Function

As before we can let the user enter their own equation. Here we set the highest power to 6. You can choose your own highest power in your program. The code for highest power of 6 is the following.

```
/*      trialimp */
/*      user enters an equation.*/
/*      equation is solved by trial & improvement */
#define _CRT_SECURE_NO_WARNINGS
#include <stdio.h>
#include <math.h>
main()
```

```
{
        float p6, p5, p4, p3, p2, p1, lower, upper;

        double testhigh, testlow, testvalue, middle;
        float constval;
        int i,  iterations, highpower;

        /* Preset constant values to zero */

        p6 = 0.0;
        p5 = 0.0;
        p4 = 0.0;
        p3 = 0.0;
        p2 = 0.0;
        p1 = 0.0;
        constval = 0.0;

        /* Enter the highest power of x in your equation */
        /* (so that you don't have to enter values if you don't have higher
        powers) */

        printf("enter highest power of x (max 6)");
        scanf("%d", &highpower);

        /* Enter the coefficient for each of your powers */

        switch (highpower) {
        case 6:
                printf("enter coefficient of x power 6(-9 to 9)");
                scanf("%f", &p6);
        case 5:
                printf("enter coefficient of x power 5(-9 to 9)");
                scanf("%f", &p5);
        case 4:
                printf("enter coefficient of x power 4(-90 to 9)");
                scanf("%f", &p4);
        case 3:
                printf("enter coefficient of x power 3(-9 to 9)");
                scanf("%f", &p3);
```

```
case 2:
      printf("enter coefficient of x power 2(-9 to +9)");
      scanf("%f", &p2);
case 1:
      printf("enter coefficient of x (-9 to +9)");
      scanf("%f", &p1);
case 0:
      printf("enter numeric constant (-999 to +999)");
      scanf("%f", &constval);
default:
      printf("default");

}
/* Display the equation you have entered */

printf("Equation is:- ");
printf("%f x**6+%f x**5+%f x**4+%f x**3 +%f x**2 %f x= %f\n", p6, p5,
p4, p3, p2, p1, constval);

/* Your curve should cross the x-axis */
/* Here, your lower value is a value of x where your curve is below
the x-axis */
/* Your upper value is a value of x where your curve is above the
x-axis */
/* Both values should be close to where your curve crosses the x-axis */

printf("enter initial lower value");
scanf("%f", &lower);
printf("enter initial upper value");
scanf("%f", &upper);
printf("enter number of iterations");
scanf("%d", &iterations);

testlow = lower;
testhigh = upper;
```

```
for (i = 0;i < iterations;i++)
      {
              middle = (testhigh + testlow) / 2;
              testvalue = p6 * pow(middle, 6) + p5 * pow(middle, 5) + p4 *
              pow(middle, 4) + p3 * pow(middle, 3) + p2 * pow(middle, 2) +
              p1 * middle + constval;
              if (testvalue == 0)
              {
                      printf("x is %f", middle);
                      return(0);
              }
              if (testvalue > 0)
              {
                      testhigh = middle;
              }
              else
              {
                      testlow = middle;
              }

      }
      printf("x is %f", middle);

}
```

EXERCISES

Solve the following equations by Trial and Improvement either by typing the formula into the code and compiling your new program as in your earlier examples or by entering it in the latter program. You will need to find the graph of the function first, so that you can find initial upper and lower points. You can do this by any graphing software you have. Graph and Autograph are possibilities. If you don't have graphing software, you can find the upper and lower values in the appendix.

1. $y = x^2 - 6x + 8$

2. $y = x^2 - 8x + 12$

3. $y = 2.6x^3 - 17.3x - 6.5$

4. $y = 5x^6 - 13x^4 + 2x$

5. $y = x^3 + 5$

6. $y = x^4 + 2x^3 - 6x^2$

7. $y = 2x^2 + x + 3$

8. Amend your program which contains the actual function to be used in the Trial and Improvement to contain a function containing inverses. The function you need to include is $y = 2x^2 - 2/x^2$.

9. Do a similar thing as question 8 but using the function $y = 0.75x^2 - 0.8x^3$.

10. Do a similar thing as question 8 but this time using the function

$y = 1(/(10x^5) - 7x^6/10$.

CHAPTER 3

Numerical Integration

In calculus if we want to find the area under a curve, there is often a simple technique of calculus for us to use. Consider the curve in Figure 3-1. This is the curve of $y = 2x - x^2$.

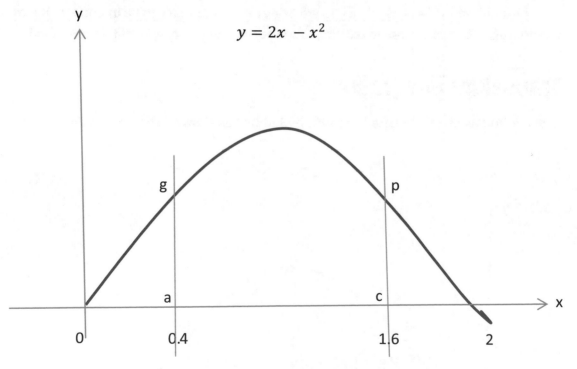

Figure 3-1. *Finding the area between a curve and x-axis*

The technique involves adding one to the power and bringing the power down to divide.

This is specified by

$$\int_{0.4}^{1.6} 2x - x^2 \, dx$$

P. Joyce, *Numerical C*, https://doi.org/10.1007/978-1-4842-5064-8_3

In this case we have chosen to find the area between 0.4 and 1.6 on the x-axis. So when we do the integration and substitute the values of x, we get the area below the curve.

1. Integrating the function $y = 2x - x^2$ (adding one to the power and dividing by that number), we get $2x^2/2 - x^3/3 + c$ where c is a constant.

2. Putting in our two limits of x as 1.6 and 0.4, we get

$$(1.6^2 - 1.6^3/3 + c) - (0.4^2 - 0.4^3/3 + c)$$

3. Evaluating this we get the area as 1.056.

This method works perfectly if you are using a function that can be integrated, but in many cases, especially those in real life, this is not so and you need a different method.

Trapezium Integration

One method used is called the "Trapezium Method," as shown in Figure 3-2.

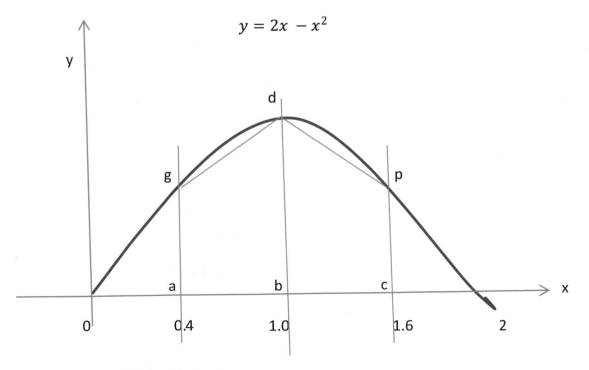

Figure 3-2. *Splitting the area into trapezia*

You may have noticed on our graph that the region of interest for our area has some extra lines drawn on it. The lines ag, gd, db, and ab mark out a trapezium, and similarly for bd, dp, pc, and bc. If we find the areas of these two trapezia, it should be fairly close to the area we have just found using calculus. It certainly looks fairly close on the diagram.

The area of a trapezium is given by the formula

Area = 0.5 x (sum of parallel sides) x perpendicular height.

In our two trapezia, the perpendicular heights are ab and bc. From the diagram you can see that the length of each of these is 0.6.

1. We can find each of the parallel sides because the formula for the curve is $y = 2x - x^2$. We can find the point g because it has the same x value as point a (0.4).

2. So if we put this value into the formula, we get $2(0.4) - (0.4)^2$. This works out to be 0.64. So this is the length of ag.

3. We can do similar things for point d and point p. In this way we can find bd and cp.

4. When we put the x values for b and c into the equation of the curve, we get 1.0 and 0.64.

5. So using these values for our two trapezia, agdb and bdpc, we get

For agdb area = 0.5 x (0.64 + 1.0) x 0.6 = 0.492

For bdpc area = 0.5 x (1.0 + 0.64) x 0.6 = 0.492

We get the same value which shows that the graph is symmetrical about the line x = 1.

6. Adding these together, we get 0.984 for the combined area of the two trapezia. This is less than our calculus answer which was 1.056, but we would expect it to be less because, as you can see from the graph, there is a gap between the top side of each of our trapezia and the curve. However, all is not lost. We can narrow this gap by splitting our area into four trapezia instead of two.

This is shown on the graph in Figure 3-3.

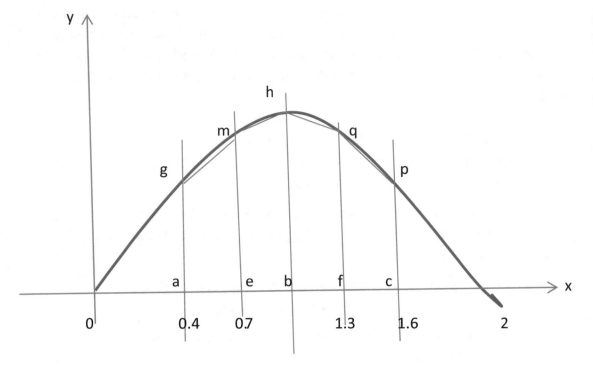

Figure 3-3. *Splitting of our area into more trapezia*

In this case our four trapezia are agme, emhb, bhqf, and fqpc.

The perpendicular height for each trapezium in this case is 0.3.

1. Again using the equation of the graph $y = 2x - x^2$, we can find the y value of g, m, h, q, and p by substituting their x values 0.4, 0.7, 1.0, 1.3, and 1.6.

2. So the corresponding y values for each of the points are 0.64, 0.91, 1.0, 0.91, 0.64.

3. So the areas of our four trapezia are

    For agme area = 0.5 x (0.64 + 0.91) x 0.3 = 0.2325

    For emhb area = 0.5 x (0.91 + 1.0) x 0.3 = 0.2865

    For bhgf area = 0.5 x (1.0 + 0.91) x 0.3 = 0.2865

    For fqpc area = 0.5 x (0.91 + 0.64) x 0.3 = 0.2325

4. So the total area of the four trapezia is 1.038.

This is closer to our calculus value of 1.056. Again you can see from the graphs why this is. The space we are missing in the areas of the trapezia is that below the curve but above each trapezium. In the second graph, this missing space is less. We could carry on splitting each trapezium into two and we will get closer to the calculus answer, but this is where our numerical C comes in. This example could be solved with calculus. We only used it in our Trapezium Method here so we could see how close we were getting to the correct answer.

Simplification of Formula

We can simplify our trapezium calculations by using a formula. Look at the graph in Figure 3-4.

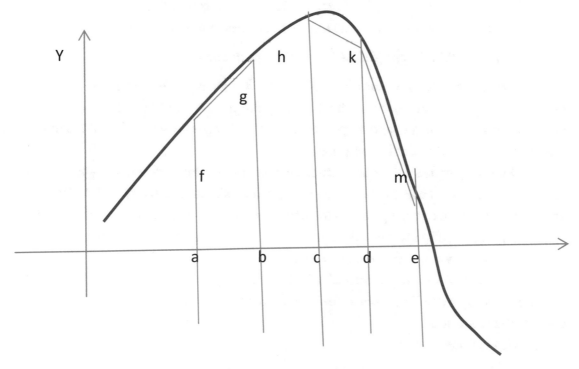

Figure 3-4. *General case for Trial and Improvement formula*

The lengths ab, bc, cd, and de are equal. We always make this the case. If we look at the calculations for the areas of our four trapezia here, this is what we should get.

1. Area = 0.5xheightx(af+bg)+0.5xheightx(bg+ch)+0.5xheightx (ch+dk)+0.5xheightx(dk+em)

2. We can tidy this up a bit as each term has 0.5xheight (where the height is the same for each trapezium).

3. We get 0.5xheightx((af+bg)+(bg+ch)+(ch+dk)+(dk+em))

4. We can even collect the terms within the outside brackets and get

Area=0.5 x height x (af+2bg+2ch+2dk+em)

You can see that for, say, the first two trapezia, bg is used for the right side of the first trapezium and the left side of the second, and so on for each pair of trapezia that are next to each other. The only sides that are not multiplied by 2 are the left side of the first trapezium and the right side of the last. We can generalize this to the case where we have any number of trapezium. The formula is

Area = 0.5 x height (first side+last side+2xall other sides)

The calculations for our examples earlier were not too bad if we have only a small number of trapezia, and if our function is a fairly simple one, but in real life the functions are a lot more complicated, and to get a high degree of accuracy which is needed in real-world cases, we, again, turn to computers.

The following code shows the calculation of the area below a graph using the Trapezium Method. It accepts a formula with powers of x up to x to power 6. Again, as in our earlier programs, it prompts the user for entry of the formula. At first, type in a function that you already know the correct answer to (possibly the function in our previous example $y = 2x - x^2$ between x=0.4 and x=1.6).

In this case it asks for the number of strips (trapezia) you want, up to a maximum of 1000.

You could try a small number at first and then a high number to see how close you get to the correct value.

This is the code.

```
/*trapezium - first attempt at trapezium method*/
#define _CRT_SECURE_NO_WARNINGS
#include <stdio.h>
#include <math.h>
main()
```

```
{

    double p6, p5, p4, p3, p2, p1, p0, lower, upper;

    double stripwidth, xposn;
    double middlevalues, zero, width, area;
    double yarr[10002]; /* Array to store lengths of sides of each
    trapezium */
    int i, strips, highpower;

    /* Preset constant values to zzero */

    p6 = 0.0;
    p5 = 0.0;
    p4 = 0.0;
    p3 = 0.0;
    p2 = 0.0;
    p1 = 0.0;
    p0 = 0.0;
    /* Enter the highest power of x in your equation */
            /* (so that you don't have to enter values if you don't have
            higher powers) */

    printf("enter highest power of x (max 6)");
    scanf("%d", &highpower);

    /* Enter the coefficient for each of your powers */
    switch (highpower) {
    case 6:
            printf("enter coefficient of x power 6(-9 to 9)");
            scanf("%lf", &p6);
    case 5:
            printf("enter coefficient of x power 5(-9 to 9)");
            scanf("%lf", &p5);
    case 4:
            printf("enter coefficient of x power 4(-90 to 9)");
            scanf("%lf", &p4);
```

```c
    case 3:
            printf("enter coefficient of x power 3(-9 to 9)");
            scanf("%lf", &p3);
    case 2:
            printf("enter coefficient of x power 2(-9 to +9)");
            scanf("%lf", &p2);
    case 1:
            printf("enter coefficient of x (-9 to +9)");
            scanf("%lf", &p1);
    case 0:
            printf("enter numeric constant (-999 to +999)");
            scanf("%lf", &p0);
    default:
            printf("default");

    }

    /* Display the equation you have entered */

    printf("Equation is:- ");
    printf("%lf x**6+%lf x**5+%lf x**4+%lf x**3 +%lf x**2 %lf x= %lf\n",
    p6, p5, p4, p3, p2, p1, p0);

    printf("enter lower limit");/* the lower x value for your
    integration */
    scanf("%lf", &lower);
    printf("enter upper limit");/* the upper x value for your integration */
    scanf("%lf", &upper);
    printf("enter number of strips (max 1000)");/* how many trapezia are
    you splitting your area into */
    scanf("%d", &strips);
    if (strips > 10000)
    {
            printf("Number of strips exceeds 10000");
            return(0);
    }
    zero = 0;
```

```
width = upper - lower;/* overall x-distance between limits of our
integration */
stripwidth = width / strips; /* stripwidth is perpendicular height of
each trapezium */
/* yarr[1] contains the First side from our trapezium area formula */
yarr[1] = p6 * pow(lower, 6) + p5 * pow(lower, 5) + p4 * pow(lower, 4)
+ p3 * pow(lower, 3) + p2 * pow(lower, 2) + (p1*lower) + p0;
/* yarr[strips+1] contains the Last side from our trapezium area
formula */
yarr[strips + 1] = p6 * pow(upper, 6) + p5 * pow(upper, 5) + p4 * pow
(upper, 4) + (p3*pow(upper, 3)) + (p2*pow(upper, 2)) + (p1*upper) + p0;
middlevalues = zero;

/* forloop loops round yarr to add all the values of the sides of the
trapezia in the formula */
for (i = 1;i < strips;i++)
{
        xposn = lower + (i*stripwidth);
        yarr[i + 1] = (p6*pow(xposn, 6) + p5 * pow(xposn, 5) + p4 *
        pow(xposn, 4) + p3 * pow(xposn, 3)) + (p2*pow(xposn, 2)) +
        (p1*xposn) + p0;
        middlevalues = middlevalues + yarr[i + 1];
}
/* Now collect the first side, the last side and 2x the middle sides and
multiply by 0.5x the stripwidth (as in the formula) */

    area = 0.5*stripwidth*(yarr[1] + 2 * middlevalues + yarr[strips + 1]);
/* Now we have the area */

    printf("Area is %lf\n", area);


}
```

The user enters the equation of their graph. Here, 6 is taken as the highest power that can be entered, but you can change this to be any power you wish. The array yarr contains the main data for the formula. yarr[1] contains the length of the left side of the first trapezium, and yarr[shapes+1] contains the right side of the last trapezium. All the other sides are multiplied by two. These are called the `middlevalues`. The `forloop` works out the length of each of these sides by taking the x coordinate of where the line meets the x-axis and inserting it into the formula of the curve. The number of iterations is specified by the user. The higher this number means, the higher the number of trapezia you use in the `forloop`. As we saw earlier, more trapezia mean that we get closer to the actual curve, therefore closer to the actual value of the area.

If you try this program using the function y= 2x – x^2 between x=0.4 and x=1.6 as in our earlier calculus example, you can see how close you get to the correct answer.

Inverse Power

It is fairly easy to add into your program the option of entering inverse powers of x for the function, for instance, if you wanted to integrate $2/x^4$. This is achieved by calling the power function (pow) with a negative number. Thus $2/x^4$ would be the same as $2x^{-4}$ so your code would be `2*pow(x,-4)` for each value of x. The following code demonstrates this. If you try entering the function y = $2/x^4$ between x limits 1 and 2, you should get an answer of 0.583.

```
/*trapezium -  trapezium method using inverse functions */
#define _CRT_SECURE_NO_WARNINGS
#include <stdio.h>
#include <math.h>
main()
{

    double p6, p5, p4, p3, p2, p1, p0, lower, upper;

    double stripwidth, xposn;
    double middlevalues, zero, width, area;
    double yarr[10002]; /* Array to store lengths of sides of each
    trapezium */
    int i, strips, highpower;

    /* Preset constant values to zzero */

    p6 = 0.0;
    p5 = 0.0;
```

```
p4 = 0.0;
p3 = 0.0;
p2 = 0.0;
p1 = 0.0;
p0 = 0.0;

/* Enter the highest inverse power of x in your equation */
/* (so that you don't have to enter values if you don't have higher
powers) */

printf("enter highest inverse power of x (max 6)");
scanf("%d", &highpower);

/* Enter the coefficient for each of your powers */

switch (highpower) {

case 6:
        printf("enter coefficient of inverse x power 6(-9 to 9)");
        scanf("%lf", &p6);
case 5:
        printf("enter coefficient of inverse x power 5(-9 to 9)");
        scanf("%lf", &p5);
case 4:
        printf("enter coefficient of inverse x power 4(-90 to 9)");
        scanf("%lf", &p4);
case 3:
        printf("enter coefficient of inverse x power 3(-9 to 9)");
        scanf("%lf", &p3);
case 2:
        printf("enter coefficient of inverse x power 2(-9 to +9)");
        scanf("%lf", &p2);
case 1:
        printf("enter coefficient of inverse x (-9 to +9)");
        scanf("%lf", &p1);
case 0:
        printf("enter numeric constant (-999 to +999)");
        scanf("%lf", &p0);
```

```
default:
        printf("default");
}

/* Display the equation you have entered */

printf("Equation is:- ");
printf("%lf x**-6+%lf x**-5+%lf x**-4+%lf x**-3 +%lf x**-2 %lf
x**-1= %lf\n", p6, p5, p4, p3, p2, p1, p0);

printf("enter lower limit");/* the lower x value for your integration */
scanf("%lf", &lower);
printf("enter upper limit");/* the upper x value for your integration */
scanf("%lf", &upper);
printf("enter number of strips (max 1000)");/* how many trapezia are
you splitting your area into */
scanf("%d", &strips);
if (strips > 10000)
{
        printf("Number of strips exceeds 10000");
        return(0);
}
zero = 0;
width = upper - lower;/* overall x-distance between limits of our
integration */
stripwidth = width / strips; /* stripwidth is perpendicular height of
each trapezium */
/* yarr[1] contains the First side from our trapezium area formula */
yarr[1] = p6 * pow(lower, -6) + p5 * pow(lower, -5) + p4 *
pow(lower, -4) + p3 * pow(lower, -3) + p2 * pow(lower, -2) +
p1 * pow(lower, -1) + p0;
/* yarr[strips+1] contains the Last side from our trapezium area
formula */
yarr[strips + 1] = p6 * pow(upper, -6) + p5 * pow(upper, -5) + p4 *
pow(upper, -4) + (p3*pow(upper, -3)) + (p2*pow(upper, -2)) + p1 *
pow(upper, -1) + p0;
middlevalues = zero;
```

```
/* forloop loops round yarr to add all the values of the sides of the
trapezia in the formula */
for (i = 1;i < strips;i++)
{
        xposn = lower + (i*stripwidth);
        yarr[i + 1] = p6 * pow(xposn, -6) + p5 * pow(xposn, -5) +
        p4 * pow(xposn, -4) + (p3*pow(xposn, -3)) +
        (p2*pow(xposn, -2)) + p1 * pow(xposn, -1) + p0;
        middlevalues = middlevalues + yarr[i + 1];
}

area = 0.5*stripwidth*(yarr[1] + 2 * middlevalues + yarr[strips + 1]);

printf("Area is %lf\n", area);

}
```

This is almost identical to the previous code except that it asks for and uses inverse powers of x.

Combined Powers

Of course you can combine the two, for instance, if you wanted to find the area under the curve $y = 3x^2 - 1/x^3$.

The code is just a mixture of the previous two programs. Here, in the interest of clarity, we restrict the maximum powers to 4 for both direct powers of x and inverse powers.

The code for this is as follows. If you try the preceding function between x–1 and x=2, you should get an answer of approximately 6.375.

```
/*trapezium -  trapezium method using a mix of direct and inverse functions */
#define _CRT_SECURE_NO_WARNINGS
#include <stdio.h>
#include <math.h>
main()
```

```
{
    double p6, p5, p4, p3, p2, p1, p0, lower, upper;
    double ip6, ip5, ip4, ip3, ip2, ip1, ip0;
    double stripwidth, xposn;
    double middlevalues, zero, width, area;
    double yarr[10002]; /* Array to store lengths of sides of each
    trapezium */
    int i, strips, highpower;

    /* Preset constant values to zzero */

    p6 = 0.0;
    p5 = 0.0;
    p4 = 0.0;
    p3 = 0.0;
    p2 = 0.0;
    p1 = 0.0;
    p0 = 0.0;
    ip6 = 0.0;
    ip5 = 0.0;
    ip4 = 0.0;
    ip3 = 0.0;
    ip2 = 0.0;
    ip1 = 0.0;
    ip0 = 0.0;

    /* Enter the highest inverse power of x in your equation */
    /* (so that you don't have to enter values if you don't have higher
    powers) */

    printf("enter highest inverse power of x (max 4)");
    scanf("%d", &highpower);

    /* Enter the coefficient for each of your powers */

    switch (highpower) {
```

```
case 4:
        printf("enter coefficient of inverse x power 4(-90 to 9)");
        scanf("%lf", &ip4);
case 3:
        printf("enter coefficient of inverse x power 3(-9 to 9)");
        scanf("%lf", &ip3);
case 2:
        printf("enter coefficient of inverse x power 2(-9 to +9)");
        scanf("%lf", &ip2);
case 1:
        printf("enter coefficient of inverse x (-9 to +9)");
        scanf("%lf", &ip1);
case 0:
        printf("enter numeric constant (-999 to +999)");
        scanf("%lf", &ip0);
default:
        printf("default");
}
/* Enter the highest power of x in your equation */
/* (so that you don't have to enter values if you don't have higher
powers) */

printf("enter highest power of x (max 4)");
scanf("%d", &highpower);

/* Enter the coefficient for each of your powers */

switch (highpower) {

case 4:
        printf("enter coefficient of x power 4(-90 to 9)");
        scanf("%lf", &p4);
case 3:
        printf("enter coefficient of x power 3(-9 to 9)");
        scanf("%lf", &p3);
```

```c
case 2:
        printf("enter coefficient of x power 2(-9 to +9)");
        scanf("%lf", &p2);
case 1:
        printf("enter coefficient of x (-9 to +9)");
        scanf("%lf", &p1);
case 0:
        printf("enter numeric constant (-999 to +999)");
        scanf("%lf", &p0);
default:
        printf("default");
}

/* Display the equation you have entered */

printf("Equation is:- ");
printf("%lf x**-4+%lf x**-3 +%lf x**-2 %lf x**-1= %lf\n", p4, p3, p2,
p1, p0);
printf("%lf x**-4+%lf x**-3 +%lf x**-2 %lf x**-1= %lf\n", ip4, ip3,
ip2, ip1, ip0);

printf("enter lower limit");/* the lower x value for your integration */
scanf("%lf", &lower);
printf("enter upper limit");/* the upper x value for your
integration */
scanf("%lf", &upper);
printf("enter number of strips (max 1000)");/* how many trapezia are
you splitting your area into */
scanf("%d", &strips);
if (strips > 10000)
{
        printf("Number of strips exceeds 10000");
        return(0);
}
zero = 0;
width = upper - lower;/* overall x-distance between limits of our
integration */
```

```
stripwidth = width / strips; /* stripwidth is perpendicular height of
each trapezium */
/* yarr[1] contains the First side from our trapezium area formula */
yarr[1] = p4 * pow(lower, 4) + p3 * pow(lower, 3) + p2 * pow(lower,
2) + p1 * pow(lower, 1) + p0 + ip4 * pow(lower, -4) + ip3 *
pow(lower, -3) + ip2 * pow(lower, -2) + ip1 * pow(lower, -1);
/* yarr[strips+1] contains the Last side from our trapezium area
formula */
yarr[strips + 1] = p4 * pow(upper, 4) + (p3*pow(upper, 3)) +
(p2*pow(upper, 2)) + p1 * pow(upper, 1) + p0 + p4 * pow(upper, -4) +
(p3*pow(upper, -3)) + (p2*pow(upper, -2)) + p1 * pow(upper, -1);
middlevalues = zero;

/* forloop loops round yarr to add all the values of the sides of the
trapezia in the formula */
for (i = 1;i < strips;i++)
{
        xposn = lower + (i*stripwidth);
        yarr[i + 1] = p4 * pow(xposn, 4) + (p3*pow(xposn, 3))
        + (p2*pow(xposn, 2)) + p1 * pow(xposn, 1) + p0 + ip4 *
        pow(xposn, -4) + (ip3*pow(xposn, -3)) + (ip2*pow(xposn, -2)) +
        ip1 * pow(xposn, -1);
        middlevalues = middlevalues + yarr[i + 1];
}

area = 0.5*stripwidth*(yarr[1] + 2 * middlevalues + yarr[strips + 1]);
printf("Area is %lf\n", area);
}
```

The flowchart in Figure 3-5 shows the overall logic of the program that can read all of the data types discussed – powers of x, inverse powers of x, and exponentials.

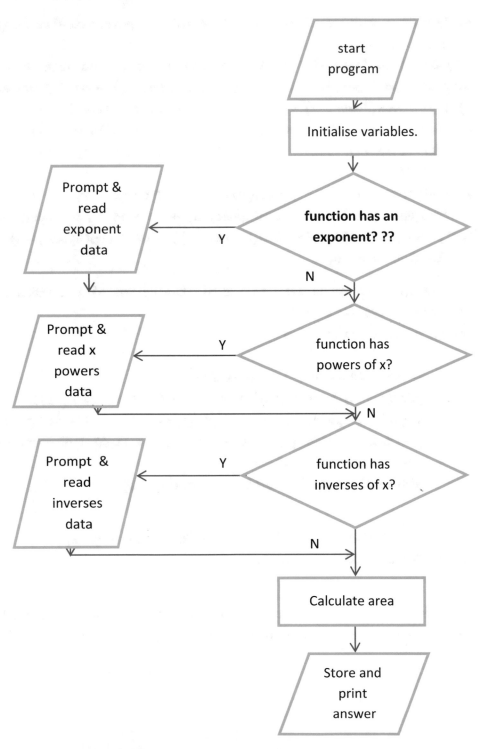

Figure 3-5. *Flowchart of logic for general method*

Problem with Negative Areas

When finding an area with integration, either using calculus or our numerical methods, we can hit a problem with negative areas. Take a look at the graph in Figure 3-6.

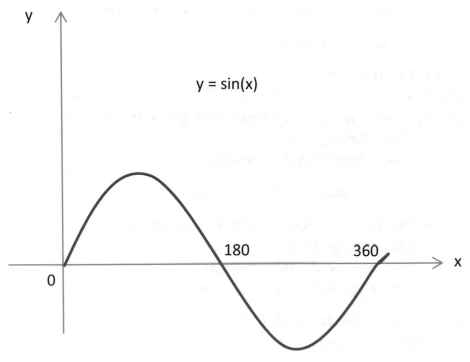

Figure 3-6. *Sine curve*

This is the sine function from 0 to 360 degrees. We can find the area between the curve and the x-axis using our Trapezium Method as before. You should be able to write this code for yourself, but it is included in the following code.

```
/*trapezium - trapezium method showing problem with negative areas */
#define _CRT_SECURE_NO_WARNINGS
#include <stdio.h>
#include <math.h>
main()
{

    double lowerx, upperx;
    double PI;
```

```c
double stripwidth, xposn;
double middlevalues, zero, width, area;
double yarr[10002];
int i, strips;

double lowangle, uppangle, lowradianno, uppradianno;

/* Preset constant values */

PI = 3.141592654;
zero = 0;
printf("Please enter lower angle in degrees:\n ");
scanf("%lf", &lowangle);
printf("You entered %f\n", lowangle);

lowradianno = lowangle * (2 * PI / 360);

printf("Please enter upper angle in degrees:\n ");
scanf("%lf", &uppangle);
printf("You entered %lf\n", uppangle);
uppradianno = uppangle * (2 * PI / 360);

printf("enter number of strips");
scanf("%d", &strips);

if (strips > 10000)
{
      printf("Number of strips exceeds 10000");
      return(0);
}
width = uppradianno - lowradianno;
stripwidth = width / strips;

lowerx = sin(lowradianno);
yarr[1] = lowerx;
upperx = sin(uppradianno);
yarr[strips + 1] = upperx;

middlevalues = zero;
```

```
    for (i = 1;i < strips;i++)
    {
            xposn = sin(lowradianno + (i*stripwidth));
            yarr[i + 1] = xposn;

            middlevalues = middlevalues + yarr[i + 1];
    }
area = 0.5*stripwidth*(yarr[1] + 2 * middlevalues + yarr[strips + 1]);

    printf("Area is %lf\n", area);

}
```

If you run this code and enter x values between 0 and 180 degrees, you should get approximately +2.0 as your area. If you rerun the program but use x values between 180 and 360 degrees, your answer will be approximately –2.0. So if you combine these and integrate between 0 and 360, you will get zero as your answer. You need to be aware of this. If it was actual physical areas you wanted, you would have to integrate the two areas separately, as we did about, but then take +2.0 rather than –2.0 for the second area. But be careful. Look at the graph in Figure 3-7.

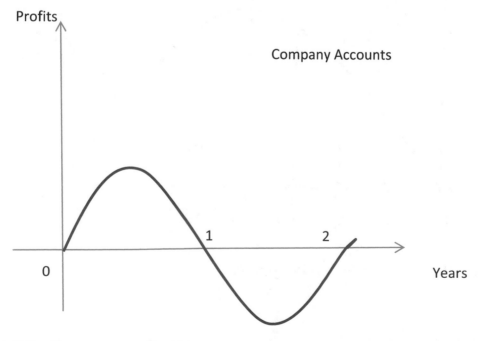

Figure 3-7. *Sine curve application*

This is basically the same graph that we have just integrated. In this case we have a real-life situation of a company's accounts over 2 years. As you can see, in the first year the company made a profit, but in the second year it made a loss. From the areas you can see that the loss in the second year was the same as the profit in the first. So, over the 2 years, the profit was zero. Thus, in this case, leaving the negative area as negative was the correct thing to do.

Simpson's Rule Integration

A method closely related to the Trapezium Method of integration is the Simpson's Rule method. Here, instead of splitting the area below the curve into trapezia, we split them into blocks where the tops are close to a parabola shape. The base and sides of each block are the same as in the Trapezium Method. Figure 3-8 illustrates this.

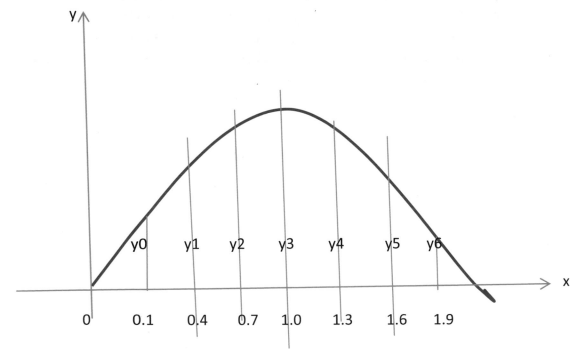

Figure 3-8. *Simpson's Rule for integration*

So really the only difference between this method and the Trapezium Method is the top of each segment. One restriction is that you must divide your area into an even number of segments. Here we have divided the area into six segments. Each segment has the same base length. Here each base will be (1.9–0.1)/6 units or 0.3 units of length. The sides of each segment (or "ordinate") are labeled y0,y1,y2,y3,y4,y5,y6.

The formula we use to find the area is

Area = (base width/3) * (1st ordinate + last ordinate + 4*odd numbered ordinates + 2*even numbered ordinates)

Again, as with the Trapezium Method, Simpson's Rule has a great advantage over calculus when the functions cannot be integrated. In our programs we just have to apply the preceding formula.

Note again that you have to split your area into an even number of sections (6 in our case earlier).

To make our programs a little simpler, we will just code the function we want to integrate, rather than asking the user to enter each element of the function. We can put a limit on how many strips we let the user split their area into. This is limited by the size of array we define to hold the formula data in the program. We can also ensure that the user enters an even number of strips. You can check this in your code. Try to work out how to do this and reject the data if it is not an even number.

Be careful that you identify the odd and even numbered ordinates. In our diagram y0 is the first, y6 is the last, the odd ones are obviously y1, y3, and y5, and the evens y2 and y4. You need to do this because, if you notice in the formula, the odd ordinates are multiplied by 4 and the even ones by 2. Work out a way you can do this in your program. The way this is done in the following program is one way. You may be able to think of a better method.

The first program using Simpson's Rule is shown here. It finds the area below the curve y = sqrt(4 + x^2) between x values 0 and 2. The function y = sqrt(4 + x^2) is coded into the program.

```c
/*simpsons - to integrate sqrt(4+x^2)*/
#define _CRT_SECURE_NO_WARNINGS
#include <stdio.h>
#include <math.h>
main()
{
    double lower, upper;
```

```c
double stripwidth, xposn;
double mideven, midodd, width, area;
double yarr[10002];
int i, strips;

/* prompt and read in limits and number of strips */

printf("enter lower limit");
scanf("%lf", &lower);
printf("enter upper limit");
scanf("%lf", &upper);
printf("enter number of strips");
scanf("%d", &strips);

if (strips > 10000)
{
    printf("Number of strips exceeds 10000");
    return(0);
}

width = upper - lower;
stripwidth = width / strips;
yarr[0] = sqrt(4 + pow(lower, 2));     /* First ordinate */
yarr[strips] = sqrt(4 + pow(upper, 2));     /* Last ordinate */

mideven = 0;
midodd = 0;
/* Process odd-numbered strips */
for (i = 1;i < strips;i++)
{
    xposn = lower + (i*stripwidth);
    yarr[i] = sqrt(4 + pow(xposn, 2));
    midodd = midodd + yarr[i];
    i++;

}
/* Process even-numbered strips */
for (i = 2;i < strips;i++)
```

```
    {

            xposn = lower + (i*stripwidth);
            yarr[i] = sqrt(4 + pow(xposn, 2));
            mideven = mideven + yarr[i];
            i++;

    }


    /* Process Simpson's formula */
    area = (0.3333)*stripwidth*((yarr[0] + yarr[strips]) + 4 * midodd + 2
* mideven);

    printf("Area is %lf\n", area);

}
```

If you copy, compile, and run this with a lower x value of 0 and an upper x value of 2 with 6 strips, you should get an answer of 4.590706.

EXERCISES

Using your Trapezium Method programs (or a program combining the different accepted functions of x), find the area under these curves between the limits shown.

1. $y = 2x^6 - x^7$ (x=1 to x=2)

2. $y = e^x$ (x=1 to x=2)

3. $y = 2 + 2x - e^x$ (x=0 to x=1)

4. $y = 2/x^4$ (x=1 to x=2)

5. $y = 2 - 1/x^2$ (x=2 to x=3)

6. $y = 3x^2 - 1/x^3$ (x=1 to x=2)

7. $y = e^{\sin x}$ (x=0 to x=PI/4)

8. $y = \exp(x^3 + 2x^2)$ (x=0.2 to x=0.6)

9. $y = e^x / (1 + e^x)$ (x=1 to x=2)

10. $y = \ln(1 + x^3)$ (x=1 to x=2)

11. $y = \sin(e^x)$ (x=0 to x=1)

12. $y = \sinh(x)$ (x=0 to x=1)

13. $y = \cosh(x)$ (x=0 to x=1)

14. $y = \tanh(x)$ (x=0 to x=1)

Using your Simpson's Rule program amended with the relevant function, find the area under these curves between the limits shown.

15. $y = \ln(x)$ (x=1 to x=4)

16. $y = \exp(-x^2)$ (x=0 to x=2)

17. $y = 1 / (1+x^4)$ (x=1 to x=2)

18. $y = \exp(x^3 + 2x^2)$ (x=−1 to x=0)

19. $y = \ln(1 + x^3)$ (x=1 to x=2)

20. $y = \sin(\exp(x))$ (x=−1 to x=1)

CHAPTER 4

Monte Carlo Integration

This chapter will show you another way to find the area under a curve that you cannot find by normal calculus integration. We can then extend the method to find volumes and even extend to higher dimensions. Although this may sound daunting, the method is similar to something you may have already seen.

Finding an Odd-Shaped Area

You may have seen examples in school mathematics, or in puzzles in magazines, where you are asked to find the area of an odd-looking shape by counting dots (as seen in Figure 4-1).

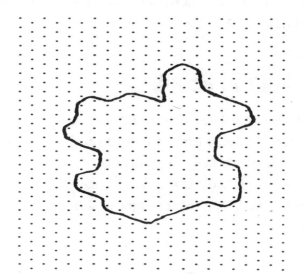

Figure 4-1. *Monte Carlo method*

© Philip Joyce 2019
P. Joyce, *Numerical C*, https://doi.org/10.1007/978-1-4842-5064-8_4

In Figure 4-1 we want to find the area of the shape inside the rectangle. If we know that the length and width of the rectangle are 7 and 6 units of length, then we can work out that the rectangle's area is 7 X 6 = 43 square units. We can then count the number of dots contained inside the rectangle.

If the number of dots is 956, then we can say that 956 dots take up 42 square units. The dots are regularly spaced throughout the rectangle. What we can do to estimate the area of our shape inside the rectangle is to count the number of dots inside the shape.

We can then say that the fraction made when you divide the number of dots in the shape by the total number of dots inside the rectangle must be the same as the fraction made when you divide the area of the shape by the area of the rectangle. If we count 379 as the number of dots inside the shape, this becomes

$$379 / 956 = (\text{area of shape}) / 42$$

Or

$$\text{Area of shape} = (42 \times 379) / 956$$

Or

$$\text{Area of shape} = 16.65 \text{ sq units}$$

If we had more dots covering the sheet, then our estimation of the area using this method would be more accurate.

Monte Carlo Area of Graph

We can use this idea to estimate the area under a graph. As with the Trapezium Method of integration, this method becomes useful when we need to find the area under a graph where the function can't be integrated using calculus.

In Figure 4-2 we are looking at the area below the curve $y = x^2$. This function can be integrated using calculus, but it is an easy example to show this technique in action.

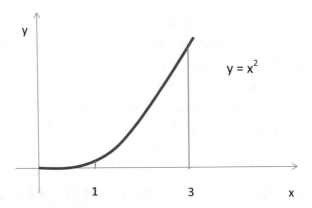

Figure 4-2. *Area between y=x² and the x-axis*

We want to find the area under the curve between x = 1 and x = 3 and the x-axis.

We create a situation similar to our previous example where we had a shape inside a dotted rectangle.

Figure 4-3 shows that we have created a rectangle abcd round the area we want to find.

The line bc rises perpendicularly from point x = 1 on the x-axis to the graph, and similarly for the line ad rising from point x = 3.

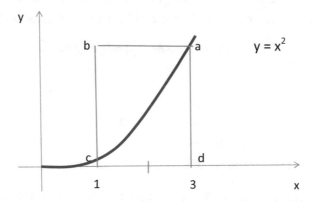

Figure 4-3. *Monte Carlo illustration*

We can now find the area of this rectangle. We know its width must be 3 – 1 = 2 units. We can find the length of ad because ad touches the graph when x = 3. So if we substitute x = 3 into the formula $y = x^2$, we get $y = 3^2 = 9$ so the length of ad must be 9. As abcd is a rectangle, then we now know that its area must be length x width = 9 x 2 = 18 square units.

We now have the shape we want to find the area of, inside a rectangle. This is close to our example of our dotted rectangle. But what about the dots? Also, it's going to be a bit annoying (and error-prone) to have to count dots.

This is where Monte Carlo comes in. Monte Carlo techniques are used in many areas of mathematics, science, and technology. Basically, they all rely on the random number generating power of computers. If you have a scientific calculator, you can get random numbers between 0 and 1 using its random number key (usually RAN or RAN# or similar).

In our rectangle earlier, we are going to use our random number generator to randomly select coordinates within the rectangle. We want to generate a random x coordinate between $x = 1$ and $x = 3$. As our C random number generator usually generates numbers between 0 and 1, we have to use a formula to convert this to numbers between 1 and 3. The formula is

$$xvalue = xlower + xrange*(rand)$$

where

xvalue is the x coordinate we want to generate

xlower is the lower value in the range we want (here 1)

xrange is the range of values of x (here 3–1=2)

and **rand** is the random number generated by the C command rand() and is between 0 and 1.

We use a similar formula to generate our random y coordinate between 0 and 9.

$$yvalue = ylower + yrange*(rand)$$

where

yvalue is the y coordinate we want to generate

ylower is the lower value in the range we want (here 0)

yrange is the range of values of x (here 9–0=9)

and **rand** is the random number generated by the C command rand() and is between 0 and 1.

So using these formulas, we generate random x coordinates between 1 and 3 and random y coordinates between 0 and 9.

When we generate the two coordinates, we add 1 to a count of points we have generated. We then test the values to see if they are below the curve. If we put the generated x value into the formula $y = x^2$, we can test if the corresponding y value is greater than our generated y value. If it is, then we know that this point is below the curve so we can add 1 to the count of points below the curve. If it is not, then we just carry on with the next generated point.

So all the points we generate are effectively the same as the dots in our original example. And, whereas we counted the dots inside our shape before, now we are counting the coordinates that we have generated which are below the curve.

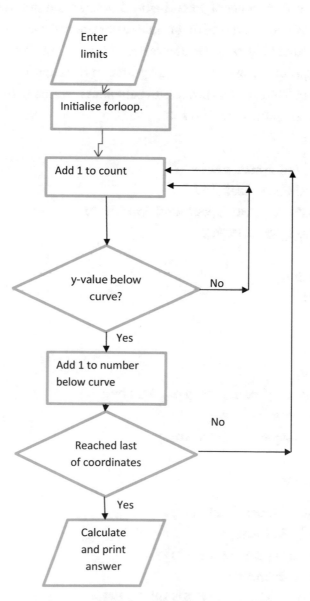

Figure 4-4. *Flowchart of logic for general method*

In Figure 4-4 the flowchart shows the logic of this method. When the program has completed its generation of points, you will have two counts. One is the total count of points generated (the same as the total number of dots on the rectangle of our original example), and the other count is the number of points below the curve (the same as the count of dots in our original shape). We can use a similar formula to our original calculation:

Area we want / Area of rectangle = points below curve / total points in rectangle

Or Area = 18 * (points below curve / total points in rectangle).

The following is C code for this process to calculate the area between the curve $y=x^2$ and the x-axis between points x=1 and x=3.

```c
/* Montecarlo */
/* integration using monte carlo */
/* by counting relative areas */
/* integrates y=x^2 to your specified limits */
#define _CRT_SECURE_NO_WARNINGS
#include <stdio.h>
#include <stdlib.h>
#include <math.h>
main()
{

    double x, y;
    double yupper, ylower, xupper, xlower;
    double  montearea, area;
    double totalexparea, totalarea;
    int j;
    int iterations;

    printf("enter lower limit\n");
    scanf("%lf", &xlower);
    printf("enter upper limit\n");
    scanf("%lf", &xupper);
    printf("xlower %lf xupper %lf\n", xlower, xupper);
    yupper = pow(xupper, 2);
    ylower = pow(xlower, 2);
    printf("ylower %lf yupper %lf\n", ylower, yupper);
    area = yupper * (xupper - xlower);
```

```c
printf("outer area is %lf\n", area);
printf("enter iterations \n");
scanf("%d", &iterations);

totalarea = 0;
totalexparea = 0;

for (j = 1;j < iterations;j++)
{

    x = rand() % 1000;/* generate random number for x up to 1000 */
    y = rand() % 1000;/* generate random number for y up to 1000 */
    y = y / 1000;/* Divide by 1000 so our number is between 0 and 1 */
    x = x / 1000;/* Divide by 1000 so our number is between 0 and 1 */
    x = xlower + (xupper - xlower)*x;/* Adjust x value to be
    between required limits */
    y = yupper * y;/* Adjust y value to be between required limits */

    if (x >= xlower)
    {
            totalarea = totalarea + 1;/* add 1 to count of points
            within whole area */

            /* test if this y value is below the curve */
            if (y <= pow(x, 2))
            {
                    totalexparea = totalexparea + 1;/* add 1 to count
                    of points below the curve */

            }
    }
}
if (totalarea != 0)
{
    montearea = area * (totalexparea / totalarea);/* calculate the
    area below the curve */
}
printf("monte area is %lf\n", montearea);

}
```

If you copy this code and compile it then run it between x lower limit of 1 and upper limit of 3, you should get a value between 8.6 and 8.7. The value if you use calculus is 8.6667.

In the program we use a technique where we call the random function rand%1000 which generates numbers up to 1000. Then we divide the number by 1000 to get values up to 1. This just gives us more accuracy in our numbers.

The following is the value you get using calculus.

$$\int_{1}^{3} x^2 \, dx$$
$$= \left[x^3 / 3 \right]_{1}^{3}$$
$$= 9 - 1/3 = 8.667$$

When you run the preceding program, you should get a value close to this (maybe 8.682).

Area of a Circle

Our next program is to find the area of a quarter of a circle. The center of the circle is the origin and it has a radius of 2 units. We will find the area in the first quadrant which is the top right quarter of the circle in Figure 4-5.

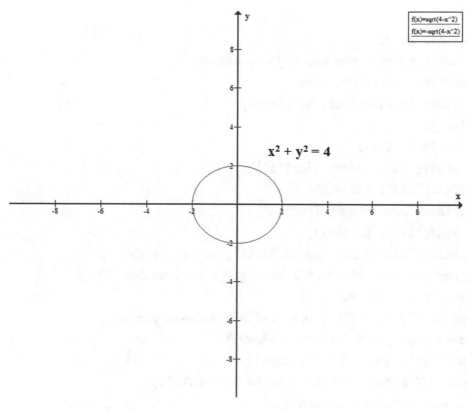

Figure 4-5. *Graph of* $x^2 + y^2 = 4$

The code for this is similar to that which we have already seen for finding the area beneath $y=x^2$. Here we want all of our x points and y points to be between 0 and 2, but the value of $x^2 + y^2$ must be less than the radius of the circle. The code for this is as follows.

```
/* Montecarlo2 */
/* integration using monte carlo */
/* by counting relative areas */
/* integrates x^2 + y^2 = 4 in the first quadrant */
#define _CRT_SECURE_NO_WARNINGS
#include <stdio.h>
#include <stdlib.h>
#include <math.h>
main()
```

```
{
        double x, y;
        double yupper, ylower, xupper, xlower;
        double  montearea, area;
        double totalexparea, totalarea;
        int j;
        int iterations;
        printf("enter lower limit\n");
        scanf("%lf", &xlower);
        printf("enter upper limit\n");
        scanf("%lf", &xupper);
        printf("xlower %lf xupper %lf\n", xlower, xupper);
        yupper = 2;/* fixed at 2 -see graph of function */
        ylower = pow(xlower, 2);
        printf("ylower %lf yupper %lf\n", ylower, yupper);
        area = yupper * (xupper - xlower);
        printf("area is %lf\n", area);
        printf("enter iterations up to 1000000\n");
        scanf("%d", &iterations);

        totalarea = 0;
        totalexparea = 0;

        for (j = 1;j < iterations;j++)
        {
                x = rand() % 1000;/* generate random number for x up to 1000 */
                y = rand() % 1000;/* generate random number for y up to 1000 */
                y = y / 1000;/* Divide by 1000 so our number is between 0 and 1 */
                x = x / 1000;/* Divide by 1000 so our number is between 0 and 1 */
                x = xlower + (xupper - xlower)*x;/* Adjust x value to be
                between required limits */
                y = yupper * y;/* Adjust y value to be between required limits */
```

```c
        if (x >= xlower)
        {
                totalarea = totalarea + 1;/* add 1 to count of points
                within whole area */
                /* test if these coordinates are within the curve */

                if (pow(x, 2) + pow(y, 2) < 4)
                {
                        totalexparea = totalexparea + 1;/* add 1 to count
                        of points below the curve */


                }
        }
}
if (totalarea != 0)
{
        montearea = area * (totalexparea / totalarea);/* calculate the
        area below the curve */
}
printf("monte area is %lf\n", montearea);

}
```

If you create this program and run it with x and y values between 0 and 2, you should get a value around 3.1645. This should be the area of our quarter circle with radius 2 units. If you use the formula for the area of a circle with radius 2 then take a quarter of this, you will get 3.1416.

We can find the area of a whole circle contained in the first quadrant (the top right quarter of the graph).

Figure 4-6 is the graph illustrating a circle of radius 2 units whose center is at coordinates (2,2).

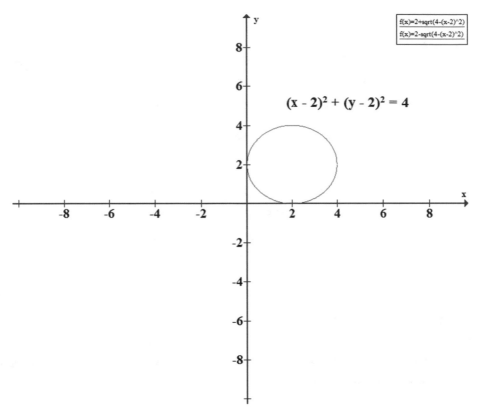

Figure 4-6. *Graph of $(x-2)^2 + (y-2)^2 = 4$*

The equation for a circle whose center is at $(2,2)$ and whose radius is 2 units is $(x-2)^2 + (y-2)^2 = 4$. We can draw a square round the circle with sides 4 units. This is our rectangle surrounding the shape whose area we need to find. We generate x values between 0 and 4 and y values between 0 and 4. We keep a count of all the coordinates we generate and a separate count of those coordinates which are enclosed by the circle. We add to the second count if the value of $(x-2)^2 + (y-2)^2$ for our coordinates (x,y) is less than 4.

The code for this is as follows.

```
/* Montecarlo circle (whole circle in 1st quadrant)*/
/* Calculation of volume using monte carlo */
/* by counting relative volumes */
/* integrates (x-2)^2 + (y-2)^2 = 2^2 to your specified limits (radius
fixed at 2 - centre at (2,2)) */
```

```c
#define _CRT_SECURE_NO_WARNINGS
#include <stdio.h>
#include <stdlib.h>
#include <math.h>
main()
{

    double x, y;
    double yupper, ylower, xupper, xlower;
    double  montearea, area;
    double totalexparea, totalarea;
    int j;
    int iterations;

    printf("enter lower x limit\n");
    scanf("%lf", &xlower);
    printf("enter upper x limit\n");
    scanf("%lf", &xupper);
    printf("xlower %lf xupper %lf\n", xlower, xupper);

    printf("enter lower y limit\n");
    scanf("%lf", &ylower);
    printf("enter upper y limit\n");
    scanf("%lf", &yupper);
    printf("ylower %lf yupper %lf\n", ylower, yupper);

    area = (xupper - xlower)*(yupper - ylower);
    printf("overall area is %lf\n", area);

    printf("enter iterations \n");
    scanf("%d", &iterations);

    totalarea = 0;
    totalexparea = 0;
    for (j = 1;j < iterations;j++)
    {
        /* find random numbers for x and y */
        x = rand() % 1000;
        y = rand() % 1000;
```

```
                y = y / 1000;
                x = x / 1000;

                /* x,y  will have numbers between 0 and 1 */
                /* so multiply by the user's entered ranges for x,y  */
                x = xlower + (xupper - xlower)*x;
                y = ylower + (yupper - ylower)*y;

                if (x >= xlower && y >= ylower)
                {
                        totalarea = totalarea + 1; /* This contains the total
                        number of entries */
                        if ((pow((y - 2), 2) + pow((x - 2), 2)) < 4)
                        {

                                totalexparea = totalexparea + 1;/* This contains
                                number of entries within desired area */

                        }
                }

        }
        if (totalarea != 0)
        {
                montearea = area * (totalexparea / totalarea);/* Monte Carlo
                area os the fraction of the outer area */
        }
        printf("monte carlo area is %lf\n", montearea);

}
```

If you run this program, you should get an area of about 12.6178. You can calculate this using a calculator and see how accurate the program is.

If you think that the Monte Carlo method is not very accurate compared to a normal mathematical approach, you would be right. BUT... the big advantage with Monte Carlo is using it when we go into higher dimensions. We will extend our calculation of the area of a circle in the first quadrant to the calculation of the volume of a sphere. As we are now in three dimensions, it won't be the first quarter we find. We can divide a 2-D graph into four sections (or quadrants), but for 3-D we divide into eight sections.

Higher Dimension Graphs

It's fairly straightforward to extend from a 2-D program to a 3-D program. We just do this same thing with z coordinates as we did with x and y coordinates in 2-D.

When we generate our 3 points (x,y,z), we then check if they are within the sphere. In this case our sphere has the same radius as our circle centered on the origin. This is 2 units. So we check if $x^2 + y^2 + z^2 < 4$ in our program.

The code for this is as follows.

```
/* Montecarlo sphere*/
/* Calculation of volume using monte carlo */
/* by counting relative volumes */
/* integrates x^2 + y^2 + z^2 = 2^2 to your specified limits (radius fixed
at 2 ) */
#define _CRT_SECURE_NO_WARNINGS
#include <stdio.h>
#include <stdlib.h>
#include <math.h>
main()
{

        double x, y, z;
        double zupper, zlower, yupper, ylower, xupper, xlower;
        double  montevol,volume;
        double totalexpvol, totalvol;
        int j;
        int iterations;

        printf("enter lower x limit\n");
        scanf("%lf", &xlower);
        printf("enter upper x limit\n");
        scanf("%lf", &xupper);
        printf("xlower %lf xupper %lf\n", xlower, xupper);

        printf("enter lower y limit\n");
        scanf("%lf", &ylower);
        printf("enter upper y limit\n");
```

```
    scanf("%lf", &yupper);
    printf("ylower %lf yupper %lf\n", ylower, yupper);

    printf("enter lower z limit\n");
    scanf("%lf", &zlower);
    printf("enter upper z limit\n");
    scanf("%lf", &zupper);
    printf("zlower %lf zupper %lf\n", zlower, zupper);

    volume = (xupper - xlower)*(yupper - ylower)*(zupper - zlower);
    printf("volume is %lf\n", volume);
    printf("enter iterations up to 1000000\n");
    scanf("%d", &iterations);

    totalvol = 0;
    totalexpvol = 0;

    for (j = 1;j < iterations;j++)
    {
        /* find random numbers for x,y and z */
        x = rand() % 1000;
        y = rand() % 1000;
        z = rand() % 1000;
        y = y / 1000;
        x = x / 1000;
        z = z / 1000;
        /* x,y and z will have numbers between 0 and 1 */
        /* so multiply by the user's entered ranges for x,y and z */
        x = xlower + (xupper - xlower)*x;
        y = ylower + (yupper - ylower)*y;
        z = zlower + (zupper - zlower)*z;

        if (x >= xlower && z >= zlower && y >= ylower)
        {
                totalvol = totalvol + 1; /* This contains the total
                number of entries */
```

```
        if ((pow(y, 2) + pow(x, 2) + pow(z, 2)) < 4)
        {

                totalexpvol = totalexpvol + 1;/* This contains
                number of entries within desired vol */

        }
    }
}
if (totalvol != 0)
{
    montevol = volume * (totalexpvol / totalvol);/* Monte Carlo
    volume os the fraction of the cube volume */
}
printf("monte carlo volume is %lf\n", montevol);

}
```

For our circle of radius 2 units, using the formula for volume, V

$$V = \frac{4}{3}\pi r^3$$

We get a value of 33.5103 for the whole sphere, but we only want one-eighth of this which is 4.1888. If you run the preceding program, your value should be fairly close to this. If you think it is not very close, then you may be right, but we have begun to move up in our dimensions where Monte Carlo can become the only method of integration.

Another 3-D example we can have a look at is a cylinder.

Here our cylinder's base is at the origin and its length moves up the y-axis.

A diagram for this is shown in Figure 4-7.

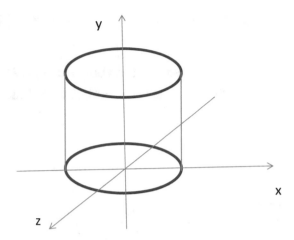

Figure 4-7. *Cylinder*

So here if the radius of the base of the cylinder is 2 units (so the diameter is 4 units) and the height is 5 units, then we can imagine our surrounding box as having dimensions 4x4x5 units.

We want to generate x and z coordinates between 0 and 4 and y coordinates between 0 and 5.

The code for this is as follows.

```
/* Montecarlo cylinder*/
/* Calculation of volume using monte carlo */
/* by counting relative volumes */
/* integrates x^2 + y^2 * z to your specified limits */
#define _CRT_SECURE_NO_WARNINGS
#include <stdio.h>
#include <stdlib.h>
#include <math.h>
main()
{

     double x, y, z;
     double zupper, zlower, yupper, ylower, xupper, xlower;
     double  montevol,  volume;
     double totalexpvol, totalvol;
     int j;
     int iterations;
```

```c
printf("enter lower x limit\n");
scanf("%lf", &xlower);
printf("enter upper x limit\n");
scanf("%lf", &xupper);
printf("xlower %lf xupper %lf\n", xlower, xupper);

printf("enter lower y limit\n");
scanf("%lf", &ylower);
printf("enter upper y limit\n");
scanf("%lf", &yupper);
printf("ylower %lf yupper %lf\n", ylower, yupper);

printf("enter lower z limit\n");
scanf("%lf", &zlower);
printf("enter upper z limit\n");
scanf("%lf", &zupper);
printf("zlower %lf zupper %lf\n", zlower, zupper);

volume = (xupper - xlower)*(yupper - ylower)*(zupper - zlower);
printf("volume is %lf\n", volume);
printf("enter iterations \n");
scanf("%d", &iterations);

totalvol = 0;
totalexpvol = 0;

for (j = 1;j < iterations;j++)
{

        /* find random numbers for x,y and z */
        x = rand() % 1000;
        y = rand() % 1000;
        z = rand() % 1000;
        y = y / 1000;
        x = x / 1000;
        z = z / 1000;
```

```
        /* x,y and z will have numbers between 0 and 1 */
        /* so multiply by the user's entered ranges for x,y and z */
        x = xlower + (xupper - xlower)*x;
        y = ylower + (yupper - ylower)*y;
        z = zlower + (zupper - zlower)*z;

        if (x >= xlower && z >= zlower && y >= ylower)
        {
                totalvol = totalvol + 1; /* This contains the total
                number of entries */

                if ((pow(y, 2) + pow(x, 2)) < 4)
                {

                        totalexpvol = totalexpvol + 1;/* This contains
                        number of entries within desired vol */


                }
            }

    }
    if (totalvol != 0)
    {
            montevol = volume * (totalexpvol / totalvol);/* Monte Carlo
            volume os the fraction of the cube volume */

    }
    printf("monte carlo volume is %lf\n", montevol);

}
```

The formula for the volume of a cylinder of base radius r and height L is

$$V = \pi r^2 L$$

Again, you can check this against the value you get when you run the preceding program. Don't forget to divide your answer from the formula by 8 as we are only finding the volume of the cylinder in the positive one-eighth section.

Following are some examples for you to work through. One example asks you to find the volume of a cone. Again, to simplify this, we will only find the positive one-eighth. The base is a circle of radius 2 units and our height is 6 units. So we enclose the cone by a box of dimensions 4x4x6. Figure 4-8 shows a cone graph.

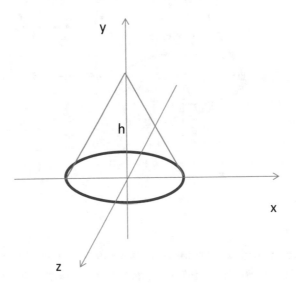

Figure 4-8. *Cone*

Our generated x and z coordinates will be between 0 and 2, and our generated y coordinate will be between 0 and 6. In this case, for a given x and z value, we want the y value to be below the slanted edge of the cone. For a given pair of x and z coordinates, we can draw a perpendicular line from the point up toward the slanted edge, and the z value must be somewhere along this line for you to add it to your inside volume count. The following diagram shows one method you can use to find this.

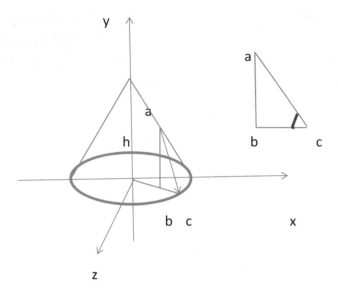

Figure 4-9. *Cone logic*

In Figure 4-9, the radius shown on the base contains our (x,z) point at b. Our perpendicular line meets the slanted edge at c. Our generated y coordinate must be along this line for it to be counted. The angle that ac makes with the base radius is always the same. The triangle shown illustrates this. If we consider a point on the circumference of the base circle, then the radius for this point makes an angle (theta) given by Tan(theta) = height/radius. We know that the height of the cone is 6 units and the base is 2 units, so tan(theta) = 6/2 = 3. This angle is the same for our triangle abc earlier.

So once we have the coordinates of point b, we can find the height ab and so we can check on this with our generated y value to see if the generated three points can be counted.

Even Higher Dimensions

We can extend our 3-D sphere program to a 4-D program. This is not entering into the field of science fiction. Even though we may not be able to envisage four dimensions of space, there is nothing odd about a 4-D graph. If we wanted to plot a graph of the variation of the pressure of a gas with volume, we can do this with a 2-D graph. If we want to add another variant to this (say temperature of the gas), we can show this on a 3-D graph. If we needed to add the magnetic field in the gas as a variant, this would be reasonable but we would not be able to show it on a graph. This type of thing occurs in all areas of life that use graphs to illustrate how one thing can vary with a number of others. For instance, in economics, we may have data from 1000 companies. For each company we have number of employees, sales value, percentage of employees who are women, and percentage of employees who are from overseas. This is four variables and we can just add one coordinate to our 3-D graph to derive information from the 4-D "graphs."

This idea of just extending our 3-D mathematics to 4-D by just adding another variable and doing the same thing with this variable as we did for the other three can also be used in computer software.

You can demonstrate this in one of the following exercises.

EXERCISES

1. Extend the example of the circle in the first quadrant to a 3-D sphere in the positive (x,y,z) sector. Check your answer with the volume formula for a sphere.

2. Complete the program to find the volume of the cone specified in the text.

3. Extend the example of the 3-D sphere in the positive (x,y,z) sector to a 4-D "sphere" in the positive (x,y,z,w) sector.

CHAPTER 5

Matrices

Matrices are mathematical structures that have become increasingly important recently. They are starting to be used in areas like data science and problems in machine learning.

Matrices can be understood using some simple examples.

Matrix Arithmetic

In Figure 5-1 we have details of three people working in a technology store selling laptops and printers. The table on the left shows details of how many of each product have been sold by each of the sales team in store and the table on the right shows how many have been sold online.

In Store			On Line		
Salesperson	Laptops	Printers	Salesperson	Laptops	Printers
Anne	3	4	Anne	10	11
Bernard	5	6	Bernard	12	13
Chris	7	8	Chris	14	15

Figure 5-1. *Matrix examples*

Each of these tables is a matrix. They both have three rows of people and two columns of items sold. So we say that each of these matrices is a 3x2 matrix (3 rows and 2 columns).

© Philip Joyce 2019
P. Joyce, *Numerical C*, https://doi.org/10.1007/978-1-4842-5064-8_5

Matrix Addition and Subtraction

If you were asked to find the totals sold for each person, it would be easy to do. You just add the laptops sold in store to the laptops sold online, similarly for printers. This is called matrix addition.

Figure 5-2 shows the matrix sum. We can do a matrix subtraction in a similar way. For instance, if we know their in-store sales and their sum, we could subtract to find their online sales.

Salesperson	Laptops	Printers
Anne	13	15
Bernard	17	19
Chris	21	23

Figure 5-2. *Matrix sum*

These examples seem very trivial but they are valuable tools. If you had a thousand staff working for the store across the country, you could detail their sales easily.

From your work with C programs up to now, you might have guessed that arrays would be useful for storing matrices.

The following is the program to add two matrices.

```
/* Matrix program */
/* Add two matrices */
#define _CRT_SECURE_NO_WARNINGS
#include<stdio.h>

main()

{
    #define MAXROW  8
    #define MAXCOL  8
    int matarr1[MAXROW][MAXCOL];/* First matrix store (rowxcolumn)*/
    int matarr2[MAXROW][MAXCOL];/* Second matrix store (rowxcolumn)*/
```

```c
int matsum[MAXROW][MAXCOL];/* Sum of matrices store (rowxcolumn)*/
int i,j,numrows,numcols;

printf("enter order of the two matrices (max 8 rows max 8 columns) \n");
scanf("%d %d", &numrows, &numcols);

/* Check if user is trying to enter too many rows or columns */

if(numrows>MAXROW || numcols>MAXCOL)
{
      printf("error - max of 8 for rows or columns\n");

}

else
{

      /* Read in first matrix */

      printf("enter first matrix\n");
      for(i=0;i<numrows;i++)
      {
            for(j=0;j<numcols;j++)
            {
                  scanf("%d",&matarr1[i][j]);
            }
      }
      printf("Your first matrix is \n");
      for(i=0;i<numrows;i++)
      {
            for(j=0;j<numcols;j++)
            {
                  printf("%d ",matarr1[i][j]);/* first matrix in
                  matarr1 */
            }
            printf("\n");
      }
```

```c
/* Read in second matrix */

printf("enter second matrix\n");
for(i=0;i<numrows;i++)
{
      for(j=0;j<numcols;j++)
      {
            scanf("%d",&matarr2[i][j]);
      }
}
printf("Your second matrix is \n");
for(i=0;i<numrows;i++)
{
      for(j=0;j<numcols;j++)
      {
            printf("%d ",matarr2[i][j]);/* second matrix in
            matarr2 */
      }
      printf("\n");
}

/* add corresponding elements of the matrices into matsum */

for(i=0;i<numrows;i++)
{
      for(j=0;j<numcols;j++)
      {
            matsum[i][j] = matarr1[i][j] + matarr2[i][j];
      }
}

/* Write the solution */

printf("Your matrix sum is \n");
for(i=0;i<numrows;i++)
{
```

```
for(j=0;j<numcols;j++)
{
        printf("%d ",matsum[i][j]);/* sum of matrices in
        matsum */
}
printf("\n");

    }
  }

}
```

You can use this program to do the matrix addition for our three workers earlier. You will have two matrices, each of three rows and two columns (3x2 matrices).

We can subtract two matrices using very similar code.

Matrix Multiplication

We can also multiply a matrix by a constant. You just need to prompt the user to enter the matrix and the constant they want to multiply by, and the code will just multiply each element of the array by the multiplier.

We can have a look at our three salespeople and we can work out the value of their sales. For this we need to know the cost of the items they sell. These details can also be held in a matrix (Figure 5-3).

Salesperson	Laptops	Printers	Cost	
Anne	13	15	Laptop	200
Bernard	17	19	Printer	25
Chris	21	23		

Figure 5-3. *Matrix multiply*

Here we are looking at their combined sales in store and online. We also have another matrix containing the price of laptops and printers. For Anne we do 13x200 and 15x25. For Bernard we do 17x200 and 19x25, and for Chris 21x200 and 23x25.

The code for this is as follows.

```c
/* Matrix program */
/* multiply two matrices */
#define _CRT_SECURE_NO_WARNINGS
#include<stdio.h>

int main()

{

    #define MAXROW  8
    #define MAXCOL  8
    int matarr1[MAXROW][MAXCOL];/* First matrix store (rowxcolumn)*/
    int matarr2[MAXROW][MAXCOL];/* Second matrix store (rowxcolumn)*/
    int matmult[MAXROW][MAXCOL];/* matrix answer (rowxcolumn)*/

    int i,j,k;
    int r1,c1,r2,c2;/* row and col for 1st and 2nd matrices */
    int error;

    error=0;

    printf("enter order of the first matrix (max 8 rows max 8 columns) \n");
    scanf("%d %d", &r1, &c1);

    /* Check if user is trying to enter too many rows or columns */
    /* or the number of columns in their 1st matrix is not the same as */
    /* the number of rows in their 2nd matrix */

    if(r1>MAXROW || c1>MAXCOL)
    {
        printf("error - max for rows or columns exceeded\n");
        error=1;

    }
```

```c
if(error == 0)
{
    printf("enter order of the second matrix (max %d rows max %d
    columns) \n",MAXROW,MAXCOL);
    scanf("%d %d", &r2, &c2);
    if(r2>MAXROW || c2>MAXCOL)
    {
        printf("error - max for rows or columns exceeded\n");
        error=1;

    }
    else
    if(c1 != r2)
    {
        printf("error - number of columns in 1st matrix must
        equal number of rows in 2nd\n");
        error=1;

    }

}
if(error == 0)
{
    for(i=0;i<r1;i++)
    {
        for(j=0;j<c2;j++)
        {
            matmult[i][j]=0;/* clear the matrix */
        }
    }

    /* Read in first matrix */

    printf("enter first matrix\n");
    for(i=0;i<r1;i++)
    {
```

```
        for(j=0;j<c1;j++)
        {
                scanf("%d",&matarr1[i][j]);
        }
}
printf("Your first matrix is \n");
for(i=0;i<r1;i++)
{
        for(j=0;j<c1;j++)
        {
                printf("%d ",matarr1[i][j]);/* first matrix in
                matarr1 */
        }
        printf("\n");
}

/* Read in second matrix */

printf("enter second matrix\n");
for(i=0;i<r2;i++)
{
        for(j=0;j<c2;j++)
        {
                scanf("%d",&matarr2[i][j]);
        }
}
printf("Your second matrix is \n");
for(i=0;i<r2;i++)
{
        for(j=0;j<c2;j++)
        {
                printf("%d ",matarr2[i][j]);/* second matrix in
                matarr2 */
        }
        printf("\n");
}
```

```
/* multiply corresponding elements of the matrices into matmult */

for(i=0;i<r1;i++)
{
    for(j=0;j<c2;j++)
    {
        for(k=0;k<r2;k++)
        {
            matmult[i][j] = matmult[i][j] + matarr1[i][k]
            * matarr2[k][j];
        }
    }
}

/* Write the solution */

printf("Your matrix multiplication is \n");
for(i=0;i<r1;i++)
{
    for(j=0;j<c2;j++)
    {
        printf("%d ",matmult[i][j]);
    }
    printf("\n");

}
    }

}
```

After the multiplication, we have another matrix (Figure 5-4).

Salesperson	Sales
Anne	2975
Bernard	3875
Chris	4775

Figure 5-4. *Resulting matrix of multiplication*

Try the code and see if you get this matrix.

So what we have done is multiplied a matrix of three rows and two columns by a matrix of two rows and one column and ended up with a matrix of three rows and one column. Or

$$(3 \times 2) \times (2 \times 1) = (3 \times 1)$$

For any matrix multiplication, the number of columns in the first matrix must equal the number of rows in the second.

It's easier to see how this works by looking at a diagram.

In Figure 5-5 we are multiplying the first two matrices to make the third. This is just another way of illustrating our computer store sales figures.

$$\begin{pmatrix} 13 & 15 \\ 17 & 19 \\ 21 & 23 \end{pmatrix} \begin{pmatrix} 200 \\ 25 \end{pmatrix} = \begin{pmatrix} 13x200 + 15x25 \\ 17x200 + 19x25 \\ 21x200 + 23x25 \end{pmatrix}$$

Figure 5-5. *Matrix multiply*

Matrix Inverse

Now that we have added, subtracted, and multiplied matrices, you may think that you should be able to divide them. There is a technique used which is a bit similar to dividing.

Consider the following equation.

$$A\{x\} = \{h\}$$

Here A, {x}, and {h} are all matrices. So we have multiplied {x} by A and we get {h} as our answer.

What if we have A and {h} and we want to find what {x} is? This is similar to the algebraic equation

$$Ax = h$$

If we had A and h here and we wanted to find x, we would just divide both sides by A.

In the matrix case, what we have to do is multiply both sides by the inverse of A. This is denoted as A^{-1}.

When you multiply both sides by this matrix, we get

$$A^{-1}A\{x\} = A^{-1}\{h\}$$

But in matrix calculations, $A^{-1}A$ is effectively 1.
So we get

$$\{x\} = A^{-1}\{h\}$$

As we know A^{-1} and $\{h\}$, we can just multiply these together using our matrix multiplication technique earlier to find $\{x\}$.

The only outstanding problem we now have is finding A^{-1}.

This method can only be used with "square matrices." A square matrix is one which has the same number of rows as columns. The easiest method is with a 2x2 matrix. The formula for finding the inverse of a 2x2 matrix is shown here.

$$\begin{pmatrix} a & b \\ c & d \end{pmatrix} = A$$

$$A^{-1} = \frac{1}{ad - bc}\begin{pmatrix} d & -b \\ -c & a \end{pmatrix}$$

For our matrix A, we see the formula for finding its inverse A^{-1}.
We can see an example of implementing this formula.

$$\text{Here A is } \begin{pmatrix} 1 & 2 \\ 3 & 4 \end{pmatrix}$$

$$\text{So } A^{-1} \text{ is} = \frac{1}{4-6}\begin{pmatrix} 4 & -2 \\ -3 & 1 \end{pmatrix}$$

$$= \frac{1}{-2}\begin{pmatrix} 4 & -2 \\ -3 & 1 \end{pmatrix}$$

$$= \begin{pmatrix} -2 & 1 \\ 3/2 & -1/2 \end{pmatrix}$$

We can test the inverse by multiplying it by the original matrix.

$$\begin{pmatrix} -2 & 1 \\ 3/2 & -1/2 \end{pmatrix}\begin{pmatrix} 1 & 2 \\ 3 & 4 \end{pmatrix} = \begin{pmatrix} 1 & 0 \\ 0 & 1 \end{pmatrix}$$

This is the answer we wanted. $\begin{pmatrix} 1 & 0 \\ 0 & 1 \end{pmatrix}$ is the "unit" 2x2 matrix. It is the equivalent of 1 in algebra.

Things get a little more complicated when we want to find the inverse of a 3x3 matrix. There are five stages:

1. Find the matrix of minors.

2. Find the matrix of cofactors.

3. Find the determinant.

4. Find the adjugate matrix.

5. Multiply the inverse of the determinant by the adjugate matrix.

1) Matrix of minors.

$$A = \begin{pmatrix} 1 & 2 & 3 \\ 3 & 2 & 1 \\ 1 & 2 & 1 \end{pmatrix}$$

Draw a line through the first row and first column. Take what you see outside of the lines and multiply the top left number by the bottom right number, then multiply the top right by the bottom left. Then subtract your second number from your first. Here we get 2x1 – 1x2 =0. So we put this answer in the first row and column position of our target matrix.

$$= \begin{pmatrix} 1 & 2 & 3 \\ 3 & 2 & 1 \\ 1 & 2 & 1 \end{pmatrix} \qquad 2x1 - 1x2 = 0 \qquad \begin{pmatrix} 0 & & \\ & & \end{pmatrix}$$

Move your lines across to the next position.

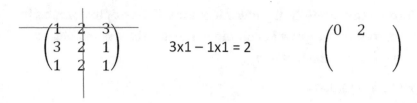

$$3x1 - 1x1 = 2$$

$$\begin{pmatrix} 0 & 2 & \\ & & \\ & & \end{pmatrix}$$

And so on.

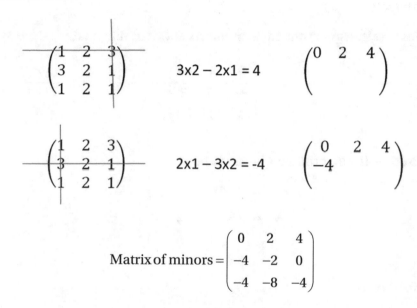

$$3x2 - 2x1 = 4$$

$$\begin{pmatrix} 0 & 2 & 4 \\ & & \\ & & \end{pmatrix}$$

$$2x1 - 3x2 = -4$$

$$\begin{pmatrix} 0 & 2 & 4 \\ -4 & & \\ & & \end{pmatrix}$$

$$\text{Matrix of minors} = \begin{pmatrix} 0 & 2 & 4 \\ -4 & -2 & 0 \\ -4 & -8 & -4 \end{pmatrix}$$

2)   Matrix of cofactors

Multiply the corresponding terms in the matrix of minors by the following array.

$$\begin{matrix} + & - & + \\ - & + & - \\ + & - & + \end{matrix}$$

Giving our matrix of cofactors

$$\begin{pmatrix} 0 & -2 & 4 \\ 4 & -2 & 0 \\ -4 & 8 & -4 \end{pmatrix}$$

3) Determinant

Here we can multiply any row or column of the original matrix by the corresponding row or column of the matrix of cofactors. So here if we take the first row

1x0 +2x(−2) + 3x4 = 8

So 8 is our determinant.

4) Adjugate

Here we transpose our cofactor matrix about its diagonal.

$$\begin{pmatrix} 0 & 4 & -4 \\ -2 & -2 & 8 \\ 4 & 0 & -4 \end{pmatrix}$$

5) Multiply the determinant by the adjugate.

$$A^{-1} = 1/8 \begin{pmatrix} 0 & 4 & -4 \\ -2 & -2 & 8 \\ 4 & 0 & -4 \end{pmatrix}$$

So A^{-1} is

$$\begin{pmatrix} 0 & 0.5 & -0.5 \\ -0.25 & -0.25 & 1 \\ 0.5 & 0 & -0.5 \end{pmatrix}$$

We can test this by multiplying it by the original matrix.

$$\begin{pmatrix} 0 & 0.5 & -0.5 \\ -0.25 & -0.25 & 1 \\ 0.5 & 0 & -0.5 \end{pmatrix} \begin{pmatrix} 1 & 2 & 3 \\ 3 & 2 & 1 \\ 1 & 2 & 1 \end{pmatrix} = \begin{pmatrix} 1 & 0 & 0 \\ 0 & 1 & 0 \\ 0 & 0 & 1 \end{pmatrix}$$

Here $\begin{pmatrix} 1 & 0 & 0 \\ 0 & 1 & 0 \\ 0 & 0 & 1 \end{pmatrix}$ is the identity matrix for a 3x3 matrix.

This is a fairly complicated procedure, but it becomes easier the more you get used to it. All you have to do is work through the five stages. You will either have to remember these or keep a note of them.

Coding a Matrix Inverse

The following is the code to invert a 3x3 matrix.

```
/* Matrix program */
/* invert a 3x3 matrix */
#define _CRT_SECURE_NO_WARNINGS
#include<stdio.h>

int main()

{

        double matarr1[3][3];/*  matrix store (rowxcolumn)*/
        double mattrans[3][3];/* adjugate matrix store (rowxcolumn)*/
        double matinv[3][3];/* matrix answer (rowxcolumn)*/

        /* array to hold the positions of the minors for each of the 9 points
        in the matrix */

        int posarr[78]={1,1,2,2,1,2,2,1,        /* row 0 col 0 */
                        1,0,2,2,1,2,2,0,        /* row 0 col 1 */
                        1,0,2,1,1,1,2,0,        /* row 0 col 2 */
                        0,1,2,2,0,2,2,1,        /* row 1 col 0 */
                        0,0,2,2,0,2,2,0,        /* row 1 col 1 */
                        0,0,2,1,0,1,2,0,        /* row 1 col 2 */
                        0,1,1,2,0,2,1,1,        /* row 2 col 0 */
                        0,0,1,2,0,2,1,0,        /* row 2 col 1 */
                        0,0,1,1,0,1,1,0};       /* row 2 col 2 */

        double det[9];/* array to contain matrix of minors row1col1,row1col2,
        row1col3,row2col1 etc */
        double detant;       /* The determinant (any row or col of the
        original matrix X corresponding one in cofactors) */
```

```c
    int i,j,x;
    int r1,c1;

    r1=3;
    c1=3;

    printf("enter matrix\n");
    for(i=0;i<r1;i++)
    {
        for(j=0;j<c1;j++)
        {
            scanf("%lf",&matarr1[i][j]);
        }
    }
    printf("Your matrix is \n");
    for(i=0;i<r1;i++)
    {
        for(j=0;j<c1;j++)
        {
            printf("%lf ",matarr1[i][j]);/* first matrix in matarr1 */
        }
            printf("\n");
    }

    /* invert */

/* Stage 1- Matrix of minors */

    for(j=0;j<9;j++)
    {
        x = j*8;
        for(i=0;i<8;i++)
        {
            det[j]=matarr1[posarr[x]][posarr[x+1]]*matarr1[posarr
            [x+2]][posarr[x+3]]-matarr1[posarr[x+4]][posarr[x+5]]*
            matarr1[posarr[x+6]][posarr[x+7]];

        }
    }
```

```
    printf("Your matrix of minors is \n");
    for(j=0;j<3;j++)
    {

        for(i=0;i<3;i++)
        {
            printf("%lf ",det[i+3*j]);
        }
        printf("\n");

    }

/* Stage 2 - Matrix of cofactors */

    printf("Your matrix of cofactors is \n");

    det[1]=det[1]*-1;
    det[3]=det[3]*-1;
    det[5]=det[5]*-1;
    det[7]=det[7]*-1;

    for(j=0;j<3;j++)
    {

        for(i=0;i<3;i++)
        {
            printf("%lf ",det[i+3*j]);
        }
        printf("\n");

    }

/* Stage 3 - Determinant */

    /* We can multiply any row or column of the original matrix by */
    /* the corresponding row or column of the matrix of cofactors. */
    /* Here we just take the first row */

    detant=matarr1[0][0]*det[0]+matarr1[0][1]*det[1]+matarr1[0][2]*det[2];
    printf("determinant is %lf ",detant);
```

```c
/* Stage 4- Adjugate (or transpose) */

    /* Transpose the cofactor matrix about its diagonal */

    printf("Your matrix transpose is  \n");
    for(j=0;j<3;j++)
    {
        for(i=0;i<3;i++)
        {
            mattrans[i][j]=det[i+3*j];

        }
        printf("\n");
    }
    for(j=0;j<3;j++)
    {
        for(i=0;i<3;i++)
        {
            printf("%lf ",mattrans[j][i]);

        }
        printf("\n");
    }

/* Stage 5- Multiply inverse of determinant by adjugate */

    /* Multiply the result of Stage 4 by the result of Stage 3 */

    for(j=0;j<3;j++)
    {
        for(i=0;i<3;i++)
        {
            matinv[j][i]=mattrans[j][i]*(1/detant);

        }
        printf("\n");
    }
```

```
/* Print solution */

    for(j=0;j<3;j++)
    {
        for(i=0;i<3;i++)
        {
            printf("%lf ",matinv[j][i]);
        }
        printf("\n");
    }

}
```

The five stages are labeled to make the code easier to follow. The array posarr contains the positions (row and column) for all of the nine minors. If you look back at the diagram of the input matrix, the first minor to be calculated is found by covering up the row and column of that position in the input matrix and then using the four numbers that are not covered up. We need to multiply the top left of these numbers by the bottom right, then multiply to top right by the bottom left and then subtract the second number from the first. In the case of the first number, its position in the matrix is row 1 column 1 (counting from 0). So our first two numbers are 1,1. The number we multiply this is at row 2 column 2 (counting from 0) so its numbers are 2,2. There are other ways of doing this, but this one is fairly easy to follow.

Testing the Code

You can test if your program has worked by multiplying it by your original matrix. When you do this, you should get

$$
\begin{pmatrix}
1 & 0 & 0 \\
0 & 1 & 0 \\
0 & 0 & 1
\end{pmatrix}
$$

If you get all zeros for your answer, it will be because the multiply array program shown at the beginning of this chapter used int arrays. You will need to change these to float arrays. Don't forget to change your printf statement when you are printing your matrix to do %f rather than %d.

EXERCISES

1. Rewrite your add two matrices program to add decimal numbers. Don't forget to change your scanf and printf instructions for floating point numbers.

2. Test question 1 with the following matrices.

a) $\begin{pmatrix} 1.5 & 0.3 & 1.7 \\ 2.6 & 0.1 & 0 \\ 3.1 & 4.0 & 1 \end{pmatrix} + \begin{pmatrix} 0.2 & 0.7 & 0.4 \\ 0.5 & 0.3 & 0 \\ 0 & 2.2 & 1.3 \end{pmatrix}$

b) $\begin{pmatrix} 1.5 & 0.3 & -1.7 \\ -2.6 & 0.1 & 0 \\ -3.1 & 4.0 & -1 \end{pmatrix}$

$+ \begin{pmatrix} -0.2 & 0.7 & 0.4 \\ -0.5 & 0.3 & -0.8 \\ 0 & 2.2 & 1.3 \end{pmatrix}$

3. Rewrite your multiply two matrices program to multiply two floating point matrices. Make the changes as outlined in the chapter.

4. Test your program from question 3 using the following matrices.

$$\begin{pmatrix} 1.5 & 0.3 & 1.7 \\ 2.6 & 0.1 & 0 \\ 3.1 & 4.0 & 1 \end{pmatrix} \begin{pmatrix} 0.2 & 0.7 & 0.4 \\ 0.5 & 0.3 & 0 \\ 0 & 2.2 & 1.3 \end{pmatrix}$$

5. Test your answer to the 2x2 inverted matrix problem in the chapter by multiplying the inverse with the original matrix. Make sure you are using the floating point program.

CHAPTER 6

Correlation and PMCC

The topics of Correlation and Product Moment Correlation Coefficient (PMCC) are related. They both concern scatter graphs. You will have seen simple scatter graphs of, say, people's height plotted against their weight or the value of a car plotted against its age.

Scatter Graphs and Correlation

In Figure 6-1 we see that, in general, as people get taller they get heavier and that as a car gets older its value decreases.

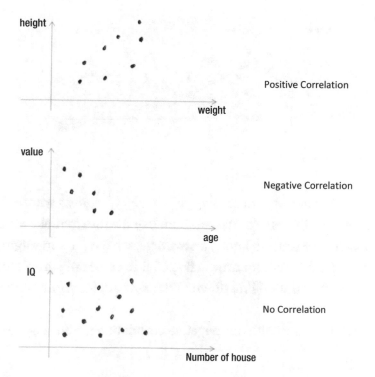

Figure 6-1. *Correlation scatter graphs*

© Philip Joyce 2019
P. Joyce, *Numerical C*, https://doi.org/10.1007/978-1-4842-5064-8_6

The first of the two graphs shows that as the x value rises, the y value rises so the graph has a positive slope. The second shows that as the x value rises, the y value falls, showing a negative slope. The first graph shows a positive correlation, and the second shows a negative correlation.

The third of the preceding graphs shows a person's IQ plotted against the number of their house or apartment. As you would expect, there is no relationship between these two so all of the points are scattered all over the graph. We say that here there is no correlation.

For the two correlation cases, we can draw a straight line showing the slopes more clearly. These are shown in Figure 6-2.

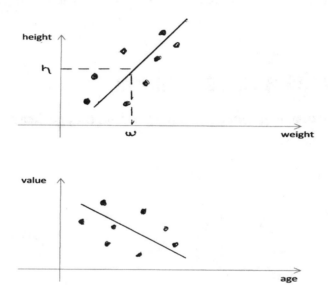

Figure 6-2. *Lines of best fit*

At school you learn to draw these "by eye," that is, you guess where the line should be. This is called a "line of best fit." We can then use the line to make an estimation. In the first graph earlier, we want to know, on average, what a person might weigh. If we know that their height is h, we can draw a dotted line across the graph to our line of best fit and then another dotted line down to the x-axis. This shows that their weight should be w.

We can do a similar thing with our negative correlation graph (Figure 6-3).

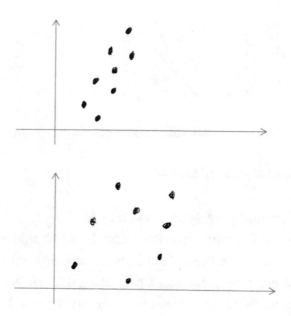

Figure 6-3. *Levels of positive correlation*

In the preceding two graphs, the first graph shows positive correlation, and it is just about reasonable to say that the second graph shows positive correlation. We say that the first graph shows "strong positive correlation" and the second shows "weak positive correlation."

So, two things are happening here. One is drawing a straight line from a scatter graph, and the other is saying how strong or weak the correlation is. Regression concerns the drawing of the line, and Product Moment Correlation Coefficient shows us how to find a fixed number to assign to the correlation rather than just using the vague terms "strong" and "weak."

The regression techniques use "least squares" techniques from statistics. The PMCC techniques give us a number from 0 to 1. In the case where all the points lie on the straight line with positive correlation, the PMCC will be +1. If all the points are on a line with a negative slope, then the PMCC is –1. In our two strong and weak positive correlation graphs, the PMCC for the strong one might be about +0.8724 and that for the weak one could be about +0.3672

If you are drawing your line of best fit by hand, a good starting point would be to find the mean of the x values and the mean of the y values and have your line going through the point of the two means. Figure 6-4 shows two possible lines of best fit. The lines cross at the means of x and y.

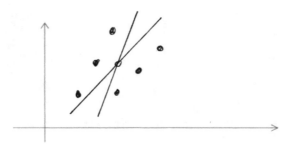

Figure 6-4. *Two possible lines of best fit*

The mean values of x and y are denoted as ($\overline{x}$, $\overline{y}$).

So for our lines of best fit in the preceding example, we should try to get as many points above our line as below it. Both of the lines we have drawn in the example do this. But there is more mathematically rigorous way of doing this. We want to try to minimize the distance of each point to our line of best fit. The two graphs in Figure 6-5 show two possible ways of doing this.

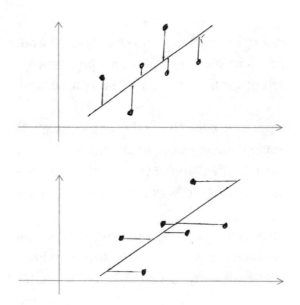

Figure 6-5. *Distances of points to the line of best fit*

The top one measures the y distance from each point to the line. This is called "regression of y on x." The lower one measures the x distance of each point to the line. This is called "regression of x on y." Both cases are trying to minimize the average distance of the points to the line. In both cases we use mathematical formulas to find the correct line.

For the x values, we use the sum of how far each of the x values is from the mean. The expression for this is

$$S_{xx} = \Sigma(x - \bar{x})^2$$

For the y values, the sum is

$$S_{yy} = \Sigma(y - \bar{y})^2$$

Also used is the sum of the x values multiplied by the y values.

$$S_{xy} = \Sigma((y - \bar{y})(x - \bar{x}))$$

There are alternative forms of these which are easier to use.

$$S_{xx} = \Sigma x^2 - (\Sigma x)^2 / n$$
$$S_{yy} = \Sigma y^2 - (\Sigma y)^2 / n$$
$$S_{xx} = \Sigma xy - (\Sigma x \Sigma y) / n$$

In these formulas n is the number of points on the scatter graph. The formula for the line of best fit for the regression of y on x case is

$$y = a + bx$$

where

$$b = S_{xy} / S_{xx} \quad a = \bar{y} - b\bar{x}$$

For the case of regression of x on y, the formula for the line is

$$x = c + dy$$

where

$$d = S_{xy} / S_{yy} \quad c = \bar{x} - d\bar{y}$$

In our program we just read all of the (x,y) values that the user types in for their scatter graph that they want to find the line of best fit for. The code puts these values into the preceding formula and then works out the equation of the line and prints it out.

The following is the code for finding the line for the regression of y on x case.

```
/* regression */
/* user enters points.*/
/* regression of y on x  calculated */
#define _CRT_SECURE_NO_WARNINGS
#include <stdio.h>
#include <math.h>
main()
{
     float xpoints[10],ypoints[10];
     float sigmax,sigmay,sigmaxy,sigmaxsquared,xbar,ybar;
     float fltcnt,sxy,sxx,b,a;
     int i,points;

     printf("enter number of points (max 10 ) \n");
     scanf("%d", &points);
     if(points>10)
     {
          printf("error - max of 10 points\n");

     }
     else
     {
          sigmax=0;
          sigmay=0;
          sigmaxy=0;
          sigmaxsquared=0;

          /* user enters points from scatter graph */
          for(i=0;i<points;i++)
          {
               printf("enter point (x and y separated by space) \n");
               scanf("%f %f", &xpoints[i], &ypoints[i]);
               sigmax=sigmax+xpoints[i];
               sigmay=sigmay+ypoints[i];
```

```c
            sigmaxy=sigmaxy+xpoints[i]*ypoints[i];
            sigmaxsquared=sigmaxsquared+(float)pow(xpoints[i],2);

        }
        printf("points are \n");
        for(i=0;i<points;i++)
        {
            printf(" \n");
            printf("%f %f", xpoints[i], ypoints[i]);

        }
        printf(" \n");
        fltcnt=(float)points;

        /* Calculation of (xbar,ybar)- the mean points*/
        /* and sxy and sxx from the formulas*/
        xbar=sigmax/fltcnt;
        ybar=sigmay/fltcnt;
        sxy=(1/fltcnt)*sigmaxy-xbar*ybar;
        sxx=(1/fltcnt)*sigmaxsquared-xbar*xbar;

        /* calculation of b and a from the formulas */
        b=sxy/sxx;
        a=ybar-b*xbar;

        /* Print the equation of the regression line */

        printf("Equation of regression line y on x  is\n ");
        printf(" y=%f + %fx", a,b);
        printf(" \n");

    }


}
```

Figure 6-6 is the graph of the regression line created by the preceding code for the points

x	1	2.2	1.4	2.8	3.2	3.4	3.8	4	5	5
y	1.4	1.6	3.2	3.8	2.8	5	4.2	6.4	5	6.4

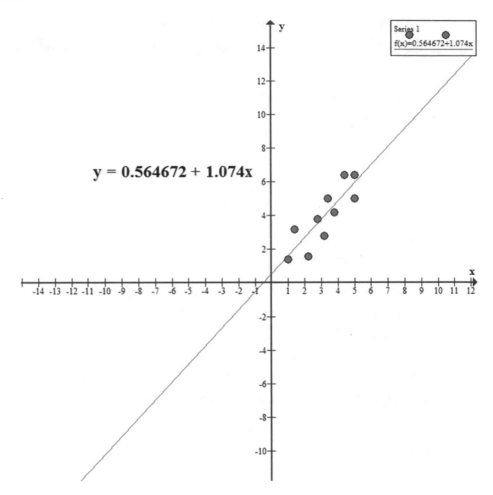

Figure 6-6. *Regression line of y on x*

The scatter points in the preceding table are shown on the graph.

The code for the regression line of x on y is given as an exercise at the end of this chapter.

Product Moment Correlation Coefficient

As we said earlier in this chapter, rather than just say that correlation is "good" or "fair" or any other term you may think of, we can assign it a number based on the statistical variables you have just used in your regression program. This number is the Product Moment Correlation Coefficient. It sounds complicated but it is just a number between

0 and 1. In this case 0 means no correlation and 1 is perfect correlation, that is, where all of the points in the scatter graph are in a straight line. We also give the number a sign, depending on whether the correlation is positive, where the regression line has a positive slope, or negative where it has a negative slope.

The formula for the PMCC is

$$r = S_{xy} / \left(S_x * S_y \right)$$

where

$$S_x = \sqrt{S_{xx}}$$

and

$$S_x = \sqrt{S_{yy}}$$

So we can just use the values of S_{xx} and S_{yy} from our regression calculations. Using the same ten scatter points as with our regression example, we can find the PMCC for that set of data.

This is a fairly minor change to the regression code. It is shown as follows.

```
/*product moment correlation coefficient - first attempt at pmcc*/
#define _CRT_SECURE_NO_WARNINGS
#include <stdio.h>
#include <math.h>
main()
{
    double xpoints[10], ypoints[10];
    double sigmax, sigmay, sigmaxsquared, sigmaysquared, xbar, ybar, sigmaxy;
    double sxy, sxx, syy, sx, sy, r;
    int i, points;
    double fltcnt;

    /* User enters number of points in scatter graph */
    printf("enter number of points (max 10 ) \n");
    scanf("%d", &points);
    if (points > 10)
```

```
        {
                printf("error - max of 10 points\n");

        }
        else
        {
                sigmax = 0;
                sigmay = 0;
                sigmaxy = 0;
                sigmaxsquared = 0;
                sigmaysquared = 0;

                /* User enters points in scatter graph */
                for (i = 0;i < points;i++)
                {
                        printf("enter point (x and y separated by space) \n");
                        scanf("%lf %lf", &xpoints[i], &ypoints[i]);
                        /* totals incremented by x and y points */
                        sigmax = sigmax + xpoints[i];
                        sigmay = sigmay + ypoints[i];
                        sigmaxy = sigmaxy + xpoints[i] * ypoints[i];
                        sigmaxsquared = sigmaxsquared + pow(xpoints[i], 2);
                        sigmaysquared = sigmaysquared + pow(ypoints[i], 2);
                }
                printf("points are \n");
                for (i = 0;i < points;i++)
                {
                        printf(" \n");
                        printf("%lf %lf", xpoints[i], ypoints[i]);

                }
                printf(" \n");
                fltcnt = points;
                /* variables in PMCC formula calculated */
                xbar = sigmax / fltcnt;
                ybar = sigmay / fltcnt;
                syy = (1 / fltcnt)*sigmaysquared - ybar * ybar;
```

```
sxx = (1 / fltcnt)*sigmaxsquared - xbar * xbar;
sx = sqrt(sxx);
sy = sqrt(syy);
sxy = (1 / fltcnt)*sigmaxy - xbar * ybar;

/* PMCC value calculated */
r = sxy / (sx*sy);
printf("r is %lf", r);
    }

}
```

If you create this program and run it with the same data as your regression program, you should get a PMCC of 0.827936.

EXERCISES

1. Starting with your program for regression of y on x, write a program to find the regression of x on y. Test it with the same data as in the y on x case.

2. If you have graph plotting software, plot the graph for the equation created by question 1.

3. Using your Product Moment Correlation Coefficient program, find the PMCC for the following 10 points on a scatter graph.

   x : 1 2 3 4 5 6 7 8 9 10

   y : 10 9 8 7 6 5 4 3 2 1

CHAPTER 7

Monte Carlo Methods

This chapter shows some of the uses of the Monte Carlo technique in mathematics, physics, and medicine. The radioactive decay simulation produces an output which matches the results from physics experiments on radioactive elements. The "Buffon's Needle" section recreates an eighteenth-century experiment to find pi. The random walk technique has applications in many walks of life particularly in physics, chemistry, and medicine.

Radioactive Decay Simulation

Some elements exhibit radioactive decay. This is where the nucleus of the element breaks up. There is a probability associated with this break up. This varies from element to element, but it is constant for a particular element. We normally denote this probability by λ. Typical values of λ are very small, and it is related to the half-life of the element. The units are time^{-1} so this means per second, per year, and so on. For Cobalt 60 the value is 0.13149 per year. This might seem very small but in a sample of, say, 1 kg of cobalt, there will be millions of nuclei so that the probability of having one decay in a short space of time would be large.

When a nucleus decays, it changes into a different nucleus so you might start with N nuclei of Cobalt 60 but after, say, 1 hour you would have less. The formula describing this is

$$N = N_o \, e^{-\lambda t}$$

where N_o is the number of nuclei at the beginning and N is the number after the unit of time t.

So if we start with a sample of Cobalt 60 of 3000 nuclei, then after 3 units of time we would have

$$N = 3000 \, e^{-0.13 \times 3}$$

which is approximately 2031.

© Philip Joyce 2019
P. Joyce, *Numerical C*, https://doi.org/10.1007/978-1-4842-5064-8_7

We can simulate this in a C program using Monte Carlo simulation. As the probability of decay is related to λt from the formula, we can generate a random number and check if it is less than λt. If it is, then the radioactive decay will take place and you will have one less Cobalt 60 nucleus in your sample. You can repeat this in a loop in your program to count how many Cobalt 60 nuclei are in the sample after each time period. Then you can use the data you produce to draw a graph. The following code shows this simulation.

```c
/* radioactive decay simulation */
#define _CRT_SECURE_NO_WARNINGS
#include <stdio.h>
#include <math.h>
#include <stdlib.h>
#include <time.h>
main()
{

    int j,timelimit,nuc;

    double randnumber,timeinc,lambda,timecount,probunittime;
    FILE *fptr;
    time_t  t;
    srand((unsigned) time(&t)); /* random number generator seed */
    fptr=fopen("radioact.dat","w");

    /* Ask user to input specific data */
    /* initial number of nuclei, the value of lambda, time for experiment */

    printf("Enter initial number of nuclei : ");
    scanf("%d",&nuc);

    printf("Enter lambda : ");
    scanf("%lf",&lambda);

    printf("Enter time   : ");
    scanf("%d",&timelimit);

    /* time increment of loop */
```

```
timeinc=0.001/lambda;

printf("Time increment :%lf",timeinc);

/* (delta t * lambda) */

probunittime=0.001*lambda;
timecount=0;

/* Monte Carlo loop */
while(timecount<=timelimit)
{
        fprintf(fptr,"%lf %d\n",timecount,nuc);

        timecount=timecount+timeinc;

        for(j=0;j<=nuc;j++)
        {
                randnumber=rand()%1000;
                randnumber=randnumber/1000;

        /* Monte Carlo method checks random number less than (delta t *
        lambda) */

                if(randnumber<=probunittime)
                        nuc=nuc-1;/* If less, then prob. that nucleus has
                        decayed */
                        if(nuc<=0)
                        goto nuclimitreached;
        }
}
nuclimitreached:    fclose(fptr); /* nuclei limit or time limit reached */

}
```

If you run this program, you are prompted for the three values. For number of nuclei, enter 100; for lambda, enter 0.13149; and for time, enter 30.

Figure 7-1 shows the number of nuclei on the y-axis and the time on the x-axis so you can see the trend of the radioactive decay. We want to write the number of nuclei existing at a particular time to a file so that we can import it into a graph using the Graph package. The file is called "radioact.dat". We open and close this file using `fopen` and `fclose`, and we write the points to the file using `fprintf`. We will see more about file access in a later chapter.

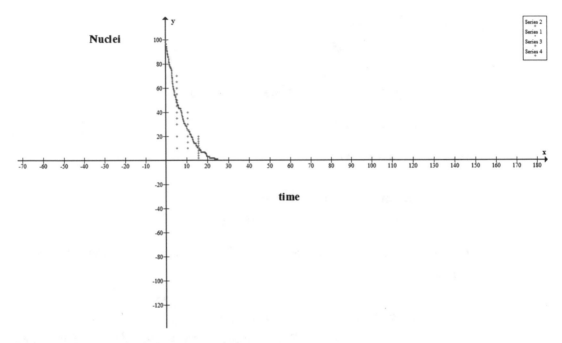

Figure 7-1. *Screenshot of radioactive decay graph*

The vertical dotted lines show the first, second, and third half-life positions, showing that the graph has a reasonable accuracy of the actual half-life values of 5.27, 10.54, and 15.81 years.

Buffon's Needle

The French aristocrat Comte de Buffon performed an interesting experiment in the eighteenth century. He took a needle of length l and dropped it. Below were two lines of space t apart. Assuming that the needle's length was smaller than the separation of the lines, there would be a probability that when you dropped the needle, it would cross one of the lines. Buffon found that this probability was

$$P = 2l / (\pi t)$$

So if you rearrange this, you get

$$\pi = 2l / (pt)$$

In 1901 Lazzarini did an experiment of this by dropping the needle onto the lines 3408 times. He found that the needle crossed one of the lines 1808 times. When he put these figures into the preceding formula, he found

$$\pi = 3.1415929$$

which is a really accurate value for π.

Rather than repeating what Lazzarini did, we can perform a similar experiment using our Monte Carlo simulation.

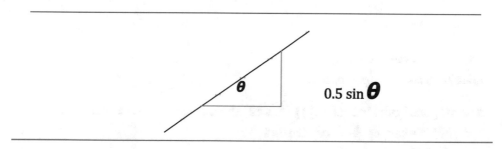

Figure 7-2. *Buffon's needle experiment*

Figure 7-2 is a diagram of the experiment. The two red lines are the parallel lines that we want to drop the needle on. The black line is the needle. For ease of calculation, we set the distance between the parallel lines to be 1 unit of length. Our diagram shows the needle after a random drop. It makes an angle of θ with the horizontal (parallel to the parallel lines). The distance of the side of our triangle opposite the angle is **0.5 sin θ**.

If the angle was a bit bigger than that in our diagram, the black needle will touch the line. We can, therefore, use this as the measure of the probability that the needle will cross one of the lines. We set the length of the needle to be 1 unit of length.

If the distance from the center of the needle to the nearest line is d, then the condition for the needle crossing the line is **d <= 0.5 sin θ**.

We can rearrange this to **2d <= sin θ**.

So we can set our two random numbers to be 2d between 0 and 0.5 and between 0 and $\pi/2$.

For our first range of numbers, this is just the same as saying d between 0 and 1.
The code for our Buffon's needle simulation is as follows.

```
/* Buffon's Needle Simulation (Monte Carlo)*/
#define _CRT_SECURE_NO_WARNINGS
#include <stdlib.h>
#include <stdio.h>
#include <math.h>
#include <time.h>

#define PI 3.141592654

main()
{
    time_t   t;

    int i, throws, count;
    double randno, anglerand;

    srand((unsigned)time(&t));/* set the random number seed */
    printf("Enter number of throws ");
    scanf("%d", &throws);

    count = 0;

    for (i = 1; i <= throws; i++)
    {
        randno = rand() % 1000;
        randno = randno / 1000;/* randno is the random number */
        anglerand = rand() % 1000;
        anglerand = anglerand / 1000;

        anglerand = 0.5*PI*anglerand; /* anglerand is the angle random
        number*/

        if (randno <= sin(anglerand))
            count = count + 1; /* Add to count */

    }
```

```
    printf("PI is %lf \n", 2 * (double)i / (double)count);
}
```

The program prints out the value of PI from the simulation.

Random Walk

The random walk is a famous method for working out many things in different areas of science. You can model the diffusion of a gas in air. It is called "a random walk" because we image a person walking along a field, say, but making turns right and left and moving forward and backward at random. The question is where would this person end up. It is sometimes described as a drunkard coming out of a bar (having indulged beyond their limits) and walking in their drunken haze in a haphazard manner. The mechanism mimics the way gas particles can move through the air. By colliding with other particles, they are bumped to the left and right or bounced backward and forward. So it is the same type of random motion as the drunkard.

So we need to know how to analyze this. If we say that the person starts off, they can make one stride forward or back or to the left or to the right. Then, after this first stride, they make another stride, again in any of the four directions and so on for many strides. If we use our random number generator, we get a number between 0 and 1. If we say that anything between 0 and 0.25 is a move to the left, between 0.25 and 0.5 is a move to the right, between 0.5 and 0.75 a move backward, and 0.75 to 1 a move forward.

Now we can just set up a loop in our program to do that sequence maybe 1000 times. We make a note of the start position (x,y), then we note whether the person walks right, left, forward, or back and add that to their position. We can then plot their position on a graph. We can just use Pythagoras' theorem to work out the position.

The following code shows this.

```
/* simple random walk simulation */
#define _CRT_SECURE_NO_WARNINGS
#include <stdio.h>
#include <stdlib.h>

#include <math.h>
#include <time.h>
```

```
FILE *output;
time_t  t;

main()
{
    int i;
    double xrand,yrand;
    double x, y, randwalkarr[10001];
    output= fopen ("randwalk4.dat", "w"); /* external file name */

    for (i=0; i<=10000; i++)
        randwalkarr [i]=0.0;              /* clear array */

    srand((unsigned) time(&t));                /* set the number generator */

    x=0.0; y=0.0;

    for (i=1;i<=10000; i++)
        {
            /* generate x random number */
            xrand=rand()%1000;
            xrand=xrand/1000;
            if(xrand<0.5)
                    x=x+1.0;
            else
                    x=x-1.0;

            /* generate y random number */
            yrand=rand()%1000;
            yrand=yrand/1000;
            if(yrand<0.5)
                    y=y+1.0;
            else
                    y=y-1.0;

            randwalkarr[i] = sqrt(x*x+y*y);/* store randwalkarr to total */
```

```
}
/* Write values to file */
for (i=0; i<=100; i++)
{

        fprintf(output,"%d %lf\n", i, randwalkarr[i*100]);
}

fclose (output);
}
```

The file read and write parts of this program will be covered in a later chapter. We can see what our random walk looks like by plotting the graph from the points collected in our loop.

This graph is shown in Figure 7-3.

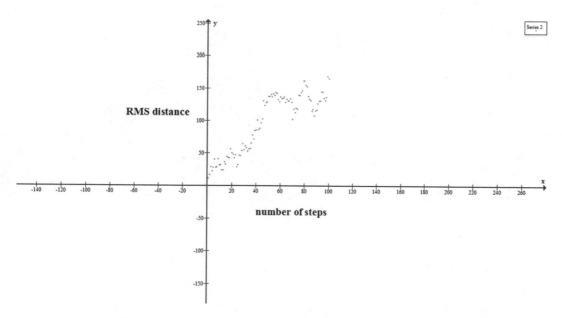

Figure 7-3. *Random walk graph*

This is the classic random walk shape. Although we have introduced this as a drunkard's walk, the mathematical nature of this process is used in many areas of science and engineering.

EXERCISES

1. Amend your random walk program to only consider a 1-D walk. So you only have to take random steps in the + or − x direction.

CHAPTER 8

Augmented Matrix

This chapter introduces a technique used in mathematics and science to solve simultaneous equations where you have many unknowns. You may be familiar with solving two simultaneous equations with two unknowns, but by the end of this chapter, you will be able to write a program to solve 12 equations with 12 unknowns.

Manual Solution to Simultaneous Equations

Solving algebraic equations is a key part of most areas of mathematics, science, technology, and many other areas. Solution of simple equations is fairly easy to do manually, but in real life the equations are usually more complicated and computer methods are of great help.

Generally, if we have one unknown quantity, then we only need one equation to solve it.

1. So for $2x + 3 = 11$

2. We write $2x = 11 - 3$

3. Or $2x = 8$

4. So $x = 4$

If we have two unknown quantities, we need two independent equations to solve them (here x+y=2 and 2(x+y)=4 are not independent equations – they are effectively the same equation).

1. So if we had

$$3x + y = 5$$

$$4x - y = 2$$

These are two independent equations with two unknowns.

© Philip Joyce 2019

P. Joyce, *Numerical C*, https://doi.org/10.1007/978-1-4842-5064-8_8

2. If we add them, we get

$7x + 0 = 7$

3. Or $x = 1$

4. So now that we know what x is, we can substitute it back into either of the original equations.

5. We get $3x1 + y = 5$

6. So $3 + y = 5$

7. So $y = 2$

8. So we have solved the equations to get $x = 1$ and $y = 2$.

If we now have three equations with three unknowns

$$2x + y - z = 1$$

$$2x - 3y + z = -1$$

$$4x - y + 4z = 14$$

This gets a little more complicated, but we can manipulate these to find x, y, and z. When we get more equations with more unknowns, the solution becomes more and more complicated and requires skill and intuition. For many equations (say 12 equations with 12 unknowns), it can become extremely difficult. The mechanism to solve these that is taught in universities is the augmented matrix method. Sometimes it has the name Row Reduction method (for reasons that will become obvious as we go on) or the Gaussian Elimination method.

If we look at the three equations earlier, we have

$$2x + y - z = 1$$

$$2x - 3y + z = -1$$

$$4x - y + 4z = 14$$

We take the coefficients of each of these equations and the numbers after the equals sign and arrange them as before.

$$\begin{array}{cccc} 2 & 1 & -1 & 1 \\ 2 & -3 & 1 & -1 \\ 4 & -1 & 4 & 14 \end{array}$$

Usually these are shown inside a matrix like this.

$$\begin{pmatrix} 2 & 1 & -1 & 1 \\ 2 & -3 & 1 & -1 \\ 4 & -1 & 4 & 14 \end{pmatrix}$$

This is called the augmented matrix.

When we solve simultaneous equations manually, we normally do things like multiply both sides of the equation by the same number or we subtract one equation from the other. We use both of these techniques in the augmented matrix technique.

We use a fixed mechanism to do this rather than relying on mathematical intuition. This technique lends itself to computational methods of solution.

We work in stages. For the solution of our three equations, we will have nine stages.

Stage 1 – We divide the first row by its first number.

Here the first number in the first row is 2 so we divide the whole row by 2, giving

$$\begin{pmatrix} 1 & 0.5 & -0.5 & 0.5 \\ 2 & -3 & 1 & -1 \\ 4 & -1 & 4 & 14 \end{pmatrix}$$

Stage 2 – We divide the second row by the first number in the second row, giving

$$\begin{pmatrix} 1 & 0.5 & -0.5 & 0.5 \\ 1 & -1.5 & 0.5 & -0.5 \\ 4 & -1 & 4 & 14 \end{pmatrix}$$

Stage 3 – We divide the third row by the first number in the third row, giving

$$\begin{pmatrix} 1 & 0.5 & -0.5 & 0.5 \\ 1 & -1.5 & 0.5 & -0.5 \\ 1 & -0.25 & 1 & 3.5 \end{pmatrix}$$

We now want to get zeroes in the first column except for the first row which we leave alone. We do this by subtracting the first row from the second row then subtracting the first row from the third row.

Stage 4 – We subtract the first row from the second row, giving

$$\begin{pmatrix} 1 & 0.5 & -0.5 & 0.5 \\ 0 & -2 & 1 & -1 \\ 1 & -0.25 & 1 & 3.5 \end{pmatrix}$$

Stage 5 – We subtract the first row from the third row, giving

$$\begin{pmatrix} 1 & 0.5 & -0.5 & 0.5 \\ 0 & -2 & 1 & -1 \\ 0 & -0.75 & 1.5 & 3 \end{pmatrix}$$

Our aim is to get a diagonal of 1's (top left to bottom right) with zeroes below the 1's. To get our next 1 in the diagonal, we divide the second row by its second number.

Stage 6 – We divide the second row by the second number in the second row, giving

$$\begin{pmatrix} 1 & 0.5 & -0.5 & 0.5 \\ 0 & 1 & -0.5 & 0.5 \\ 0 & -0.75 & 1.5 & 3 \end{pmatrix}$$

Then the same thing is done with the third row.

Stage 7 – We divide the third row by second number in the third row, giving

$$\begin{pmatrix} 1 & 0.5 & -0.5 & 0.5 \\ 0 & 1 & -0.5 & 0.5 \\ 0 & 1 & -2 & -4 \end{pmatrix}$$

We can then subtract to get the zero below the diagonal.

Stage 8 – We subtract the second row from the third row, giving

$$\begin{pmatrix} 1 & 0.5 & -0.5 & 0.5 \\ 0 & 1 & -0.5 & 0.5 \\ 0 & 0 & -1.5 & -4.5 \end{pmatrix}$$

Finally we can divide the third row by its third element.

Stage 9 – We divide the third row by its third element.

$$\begin{pmatrix} 1 & 0.5 & -0.5 & 0.5 \\ 0 & 1 & -0.5 & 0.5 \\ 0 & 0 & 1 & 3 \end{pmatrix}$$

If we now remember, our augmented matrix just represents our original three equations we have transformed.

$$\begin{pmatrix} 2 & 1 & -1 & 1 \\ 2 & -3 & 1 & -1 \\ 4 & -1 & 4 & 14 \end{pmatrix}$$

To

$$\begin{pmatrix} 1 & 0.5 & -0.5 & 0.5 \\ 0 & 1 & -0.5 & 0.5 \\ 0 & 0 & 1 & 3 \end{pmatrix}$$

So we can rewrite our original equations as

$$x + 0.5y - 0.5z = 0.5$$

$$y - 0.5z = 0.5$$

$$z = 3$$

We already have one solution, $z = 3$. We can get y by substituting $z = 3$ into the second equation to get

$$y - 3(0.5) = 0.5$$

$$\text{or } y = 2$$

then substitute $z = 3$ and $y = 2$ into the first equation to get

$$x - 2(0.5) - 3(0.5) = 0.5$$

$$\text{or } x = 0.5 - 1 + 1.5 = 1$$

So our solutions to the original three equations are

$$x = 1, y = 2, z = 3$$

Augmented Matrix Program

The nine-stage mechanism is used in our first program. We will use the original three equations from the previous section.

So the augmented matrix is our original 3x4 augmented matrix which is

$$\begin{pmatrix} 2 & 1 & -1 & 1 \\ 2 & -3 & 1 & -1 \\ 4 & -1 & 4 & 14 \end{pmatrix}$$

We will preset this matrix in a 3x4 array called matrix[3][4].

Then we proceed to perform the nine stages as described.

The code for this is as follows.

```
/* augmat2 */
/* augmented matrix 3x4 */
/* uses nine row operations for a 3 equation problem */
#define _CRT_SECURE_NO_WARNINGS
#include <stdio.h>
#include <math.h>
main()
{

    float value, x, y, z;
    float matrix[3][4];
    float divisor;
    int i, j;

/* Augmented matrix to be input is preset */

    float matrix[3][4] = {
            {2,1,-1,1},
            {2,-3,1,-1},
            {4,-1,4,14}
            };
```

```
/* Print out the preset augmented matrix */

printf("augmented matrix is\n");
for (j = 0;j < 3;j++)
{
        for (i = 0;i < 4;i++)
        {
                printf("matrix[%d][%d] = %f\n", j, i, matrix[j][i]);

        }
}
```

We work in stages. For the solution of our three equations, we will have nine stages.
Stage 1 – We divide the first row by its first number.

Here the first number in the first row is 2 so we divide the whole row by 2, so starting with

$$\begin{pmatrix} 2 & 1 & -1 & 1 \\ 2 & -3 & 1 & -1 \\ 4 & -1 & 4 & 14 \end{pmatrix}$$

We get

$$\begin{pmatrix} 1 & 0.5 & -0.5 & 0.5 \\ 2 & -3 & 1 & -1 \\ 4 & -1 & 4 & 14 \end{pmatrix}$$

The code continues as follows.

```
/* Nine stages on the rows of the augmented matrix for our 3 equation
problem */

        /* stage 1 divide first row by (row 0 col 0 ) */

divisor = matrix[0][0];
for (i = 0;i < 4;i++)
{
        matrix[0][i] = (matrix[0][i]) / divisor;
}
```

```
printf("augmented matrix after division of first row is\n");
for (j = 0;j < 3;j++)
{
      printf("%f %f %f %f\n", matrix[j][0], matrix[j][1],
      matrix[j][2], matrix[j][3]);
}
```

Stage 2 – We divide the second row by the first number in the second row, so starting with

$$\begin{pmatrix} 1 & 0.5 & -0.5 & 0.5 \\ 2 & -3 & 1 & -1 \\ 4 & -1 & 4 & 14 \end{pmatrix}$$

We get

$$\begin{pmatrix} 1 & 0.5 & -0.5 & 0.5 \\ 1 & -1.5 & 0.5 & -0.5 \\ 4 & -1 & 4 & 14 \end{pmatrix}$$

Continuing the code as follows.

```
/* stage 2 divide second row by (row 1 col 0 )*/

divisor = matrix[1][0];
for (i = 0;i < 4;i++)
{
      matrix[1][i] = (matrix[1][i]) / divisor;
}
printf("augmented matrix after division of second row is\n");
for (j = 0;j < 3;j++)
{
      printf("%f %f %f %f\n", matrix[j][0], matrix[j][1], matrix[j]
      [2], matrix[j][3]);
}
```

Stage 3 – We divide the third row by the first number in the third row, so starting with

$$\begin{pmatrix} 1 & 0.5 & -0.5 & 0.5 \\ 1 & -1.5 & 0.5 & -0.5 \\ 4 & -1 & 4 & 14 \end{pmatrix}$$

We get

$$\begin{pmatrix} 1 & 0.5 & -0.5 & 0.5 \\ 1 & -1.5 & 0.5 & -0.5 \\ 1 & -0.25 & 1 & 3.5 \end{pmatrix}$$

Continuing the code as follows.

```
/* stage 3 divide third row by (row 2 col 0 )*/
divisor = matrix[2][0];
for (i = 0;i < 4;i++)
{
    matrix[2][i] = (matrix[2][i]) / divisor;
}
printf("augmented matrix after division of third row is\n");
for (j = 0;j < 3;j++)
{
    printf("%f %f %f %f\n", matrix[j][0], matrix[j][1],
    matrix[j][2], matrix[j][3]);
}
```

We now want to get zeroes in the first column except for the first row which we leave alone. We do this by subtracting the first row from the second row then subtracting the first row from the third row.

Stage 4 – We subtract the first row from the second row, so starting with

$$\begin{pmatrix} 1 & 0.5 & -0.5 & 0.5 \\ 1 & -1.5 & 0.5 & -0.5 \\ 1 & -0.25 & 1 & 3.5 \end{pmatrix}$$

We get

$$\begin{pmatrix} 1 & 0.5 & -0.5 & 0.5 \\ 0 & -2 & 1 & -1 \\ 1 & -0.25 & 1 & 3.5 \end{pmatrix}$$

Continuing the code as follows.

```
/* stage 4 subtract first row from second row */
    divisor = matrix[1][0];
    for (i = 0;i < 4;i++)
    {
        matrix[1][i] = (matrix[1][i]) - matrix[0][i];
    }
    printf("augmented matrix after subtraction of first row from second
    row is\n");
    for (j = 0;j < 3;j++)
    {
        printf("%f %f %f %f\n", matrix[j][0], matrix[j][1], m
        atrix[j][2], matrix[j][3]);
    }
```

Stage 5 – We subtract the first row from the third row, so starting with

$$\begin{pmatrix} 1 & 0.5 & -0.5 & 0.5 \\ 0 & -2 & 1 & -1 \\ 1 & -0.25 & 1 & 3.5 \end{pmatrix}$$

We get

$$\begin{pmatrix} 1 & 0.5 & -0.5 & 0.5 \\ 0 & -2 & 1 & -1 \\ 0 & -0.75 & 1.5 & 3 \end{pmatrix}$$

Continuing the code as follows.

```
/* stage 5 subtract first row from third row*/
    divisor = matrix[2][0];
    for (i = 0;i < 4;i++)
    {
        matrix[2][i] = (matrix[2][i]) - matrix[0][i];
    }
    printf("augmented matrix after subtraction of first row from third
    row is\n");
    for (j = 0;j < 3;j++)
    {
        printf("%f %f %f %f\n", matrix[j][0], matrix[j][1], matrix[j]
        [2], matrix[j][3]);
    }
```

Our aim is to get a diagonal of 1's (top left to bottom right) with zeroes below the 1's. To get our next 1 in the diagonal, we divide the second row by its second number.

Stage 6 – We divide the second row by the second number in the second row, so starting with

$$\begin{pmatrix} 1 & 0.5 & -0.5 & 0.5 \\ 0 & -2 & 1 & -1 \\ 0 & -0.75 & 1.5 & 3 \end{pmatrix}$$

We get

$$\begin{pmatrix} 1 & 0.5 & -0.5 & 0.5 \\ 0 & 1 & -0.5 & 0.5 \\ 0 & -0.75 & 1.5 & 3 \end{pmatrix}$$

Continuing the code as follows.

```
/* stage 6 divide second row by (row 1 col 1 )*/
    divisor = matrix[1][1];
    for (i = 0;i < 4;i++)
    {
```

```
        matrix[1][i] = (matrix[1][i]) / divisor;
    }
    printf("augmented matrix after division of second row is\n");
    for (j = 0;j < 3;j++)
    {
        printf("%f %f %f %f\n", matrix[j][0], matrix[j][1], matrix[j][2],
        matrix[j][3]);
    }
```

Then the same thing is done with the third row.

Stage 7 – We divide the third row by the second number in the third row, so starting with

$$\begin{pmatrix} 1 & 0.5 & -0.5 & 0.5 \\ 0 & 1 & -0.5 & 0.5 \\ 0 & -0.75 & 1.5 & 3 \end{pmatrix}$$

We get

$$\begin{pmatrix} 1 & 0.5 & -0.5 & 0.5 \\ 0 & 1 & -0.5 & 0.5 \\ 0 & 1 & -2 & -4 \end{pmatrix}$$

Continuing the code as follows.

```
/* stage 7 divide third row by (row 2 col 1) */

    divisor = matrix[2][1];
    for (i = 0;i < 4;i++)
    {
        matrix[2][i] = (matrix[2][i]) / divisor;
    }
    printf("augmented matrix after division of third row is\n");
    for (j = 0;j < 3;j++)
    {
        printf("%f %f %f %f\n", matrix[j][0], matrix[j][1],
        matrix[j][2], matrix[j][3]);
    }
```

We can then subtract to get the zero below the diagonal.

Stage 8 – We subtract the second row from the third row, so starting with

$$\begin{pmatrix} 1 & 0.5 & -0.5 & 0.5 \\ 0 & 1 & -0.5 & 0.5 \\ 0 & 1 & -2 & -4 \end{pmatrix}$$

We get

$$\begin{pmatrix} 1 & 0.5 & -0.5 & 0.5 \\ 0 & 1 & -0.5 & 0.5 \\ 0 & 0 & -1.5 & -4.5 \end{pmatrix}$$

Continuing the code as follows.

```
/* stage 8 subtract second row from third row*/
    divisor = matrix[2][0];
    for (i = 0;i < 4;i++)
    {
        matrix[2][i] = (matrix[2][i]) - matrix[1][i];
    }
    printf("augmented matrix after subtraction of second row from third
    row is\n");
    for (j = 0;j < 3;j++)
    {
        printf("%f %f %f %f\n", matrix[j][0], matrix[j][1],
        matrix[j][2], matrix[j][3]);
    }
```

Finally we can divide the third row by its third element.

Stage 9 – We divide the third row by its third element, so starting with

$$\begin{pmatrix} 1 & 0.5 & -0.5 & 0.5 \\ 0 & 1 & -0.5 & 0.5 \\ 0 & 0 & -1.5 & -4.5 \end{pmatrix}$$

We get

$$\begin{pmatrix} 1 & 0.5 & -0.5 & 0.5 \\ 0 & 1 & -0.5 & 0.5 \\ 0 & 0 & 1 & 3 \end{pmatrix}$$

Continuing the code as follows.

```
/* stage 9 divide third row by (row 2 col 2 )*/

    divisor = matrix[2][2];
    for (i = 0;i < 4;i++)
    {
        matrix[2][i] = (matrix[2][i]) / divisor;
    }
    printf("augmented matrix after division of third row is\n");
    for (j = 0;j < 3;j++)
    {
        printf("%f %f %f %f\n", matrix[j][0], matrix[j][1],
        matrix[j][2], matrix[j][3]);
    }
```

If we now remember, our augmented matrix just represents our original three equations we have transformed.

$$\begin{pmatrix} 2 & 1 & -1 & 1 \\ 2 & -3 & -1 & -1 \\ 4 & -1 & 4 & 14 \end{pmatrix}$$

To

$$\begin{pmatrix} 1 & 0.5 & -0.5 & 0.5 \\ 0 & 1 & -0.5 & 0.5 \\ 0 & 0 & 1 & 3 \end{pmatrix}$$

So we can rewrite our original equations as

$$x + 0.5y - 0.5\,z = 0.5$$

$$y - 0.5z = 0.5$$

$$z = 3$$

Continuing the code as follows.

```
/* Print out x,y and z solutions */

z = matrix[2][3];
y = matrix[1][3] - z * matrix[1][2];
x = matrix[0][3] - y * matrix[0][1] - z * matrix[0][2];

printf("x = %f y= %f z = %f", x, y, z);
}
```

If you create this program and run it, you should get the same solutions as we did manually. The program prints out the augmented matrix after each of the nine stages so that you can check that it is doing the correct manipulation.

As our divide and subtract stages are similar, we can write a separate function for each of these and call them as required instead of writing separate code for each stage.

The next program shows these procedures (funcdivide and funcsubtract). The program also allows you to enter the augmented matrix yourself rather than it being preset. The program prints out the augmented matrix at various points in its operation. This can be commented out if you wish.

```
/* augmat17A */
/* augmented matrix 3x4 */
/* complete program   */
/*   */
#define _CRT_SECURE_NO_WARNINGS
#include <stdio.h>
#include <math.h>

void funcdivide(int first, int second, int count);
void funcsubtract(int first, int second, int count);
```

```
double matrix[12][13];
double divisor;
int i;

main()
{

      double element, x, y, z;
      int i, j, n;

      n = 3; /* Only 3x3 square matrix for this program */
      printf("square matrix is %d", n);

/* Enter your own 3x4 augmented matrix */

      printf("Enter data for augmented matrix \n");

      for (j = 0;j < n;j++)
      {
          printf("row %d ", j);
          for (i = 0;i < n + 1;i++)
          {
                  printf("enter x\n");
                  scanf("%lf", &element);
                  matrix[j][i] = element;
          }
      }
      printf("augmented matrix is\n");
      for (j = 0;j < n;j++)
      {

          for (i = 0;i < n + 1;i++)
          {
                  printf("matrix[%d][%d] = %lf\n", j, i, matrix[j][i]);

          }
      }
```

```
/*      Perform 9 stages to find row-reduced form of augmented
        matrix.*/
/*      divide stages are done in funcdivide */
/*      subtract stages are done in funcsubtract */
```

Our divide and subtract functions are called in the following code. For the divide function (funcdivide), there are three parameters in the call to the function. The first specifies the row number where the division is done. The first and second parameters together specify the row and column numbers which tell you where the divisor is. The last parameter is the stage number of our nine stages, and this can be used to identify the stage if you print out the results of each stage during testing. So for a call to funcdivide of funcdivide(0,1,2) – this would mean that we are doing the division on row 0 (first parameter) using the divisor in row 0 column 1 (first and second parameters), and the stage is stage 2 (third parameter).

For the subtract function (funcsubtract), the first parameter is the row we are subtracting from, the second is the row we are subtracting, and the third is the stage number of our nine stages. So for a call of funcsubtract(0,1,2), we would be subtracting row 1 from row 0 and it would be stage 2.

The code for this program continues as follows.

```
/* For the funcsubtract function ...... */
/* the first and second parameters refer to the rows which are used */

/* For the funcdivide function ...... */
/* the first and second parameters refer to row and column for the divisor
*/

/* The third parameter is the Stage in our Stage 1-9 method */
/* The Stage number can be printed at the end of the function */
/* so that you can monitor the progress */

    funcdivide(0, 0, 1);        /* stage 1 - divide row 0 by divisor row 0
                                   col 0 */
    funcdivide(1, 0, 2);        /* stage 2 - divide row 1 by divisor row 1
                                   col 0 */
    funcdivide(2, 0, 3);        /* stage 3 - divide row 2 by divisor row 2
                                   col 0 */
    funcsubtract(1, 0, 4);      /* stage 4 - subtract row 0 from row 1 */
```

```
    funcsubtract(2, 0, 5);      /* stage 5 - subtract row 0 from row 2 */
    funcdivide(1, 1, 6);        /* stage 6 - divide row 1 by divisor row 1
                                   col 1 */

    funcdivide(2, 1, 7);        /* stage 7 - divide row 2 by divisor row 2
                                   col 1 */

    funcsubtract(2, 1, 8);      /* stage 8 - subtract row 1 from row 2 */
    funcdivide(2, 2, 9);        /* stage 9 - divide row 2 by divisor row 2
                                   col 2 */

    /* Calculate and print out the answers */

    z = matrix[2][3];
    y = matrix[1][3] - z * matrix[1][2];
    x = matrix[0][3] - y * matrix[0][1] - z * matrix[0][2];

    printf("x = %f y= %f z = %f", x, y, z);

}

void funcdivide(int first, int second, int count)
{
    /* divide each element in row "first" by element [first][second] */
    int i, j;

    divisor = matrix[first][second];
    for (i = 0;i < 4;i++)
    {
        matrix[first][i] = matrix[first][i] / divisor;
    }
    printf("augmented matrix after %d operation is\n", count);
    for (j = 0;j < 3;j++)
    {
        printf("%lf %lf %lf %lf\n", matrix[j][0], matrix[j][1],
        matrix[j][2], matrix[j][3]);
    }
}
```

```c
void funcsubtract(int first, int second, int count)
{
        /* subtract row "second" from row "first" */
        int i, j;


        for (i = 0;i < 4;i++)
        {
                matrix[first][i] = (matrix[first][i]) - matrix[second][i];
        }
        printf("augmented matrix after %d operation is\n", count);
        for (j = 0;j < 3;j++)
        {
                printf("%lf %lf %lf %lf\n", matrix[j][0], matrix[j][1],
                matrix[j][2], matrix[j][3]);
        }
}
```

The last two programs worked with a 3x4 augmented matrix. In the following program, you can enter a 3x4, 4x5, or 5x6 augmented matrix. You are prompted for which of these you are going to enter.

The program also sets up forloops to call the funcdivide and funcsubtract functions. In the case of 4x5 and 5x6 matrices, there are more than nine stages of division and subtraction. Here, we use forloops with limits based on the row and column values of the augmented matrix.

```c
/* augmat18 */
/* augmented matrix for 3,4 or 5 equations */
/* calls functions for row division and row subtraction */

#define _CRT_SECURE_NO_WARNINGS
#include <stdio.h>
#include <math.h>

void funcdivide(int first, int second, int count);
void funcsubtract(int first, int second, int count);

double matrix[12][13];
```

```
double divisor;
int i,j, row, col;

main()
{
    double value, x, y, z, a, b, c, d, e;

    int i,  k, n, count;

    /* Enter data for augmented matrix - one element at a time*/

    printf("enter row/column number (square matrix 3x3 4x4 5x5 only)");
    scanf("%d", &n);

    printf("square matrix is %d", n);

    row = n;
    col = n + 1;

    for (j = 0;j < n;j++)
    {
        printf("row %d ", j);
        for (i = 0;i < n + 1;i++)
        {
            printf("enter x\n");
            scanf("%lf", &value);
            matrix[j][i] = value;
        }
    }

    /* Print out the entered matrix */

    printf("augmented matrix is\n");
    for (j = 0;j < n;j++)
    {
```

```
for (i = 0;i < n + 1;i++)
{
        printf("matrix[%d][%d] = %lf\n", j, i, matrix[j][i]);

}
}
```

For our 3x4 augmented matrix, we have nine stages for division and subtraction on our matrix. For a 4x5 augmented matrix, we would have 16 stages, and for a 5x6 augmented matrix, we would have 25 stages. Rather than have a list of all of these calls like in our previous program, we can call them from two nested forloops. This is shown in the following code. You can check that the correct number is being called by substituting the appropriate values for row and col. The code can be tested for any order of matrix, for example, 6x7, 7x8, and so on, up to a maximum size of the variable "matrix" that you have defined in the program. The maximum we use here is 12x13, although if you want to solve more than 12 simultaneous equations, you can just amend your definition of "matrix" to allow this.

The code continues as follows.

```
/* Perform  stages to find row-reduced form of augmented matrix.*/
/* divide stages are done in funcdivide */
/* subtract stages are done in funcsubtract */

count = 0;
for (i = 0;i < col;i++)
{
    for (k = i;k < row;k++)
    {
        count = count + 1;
        funcdivide(k, i, count);
    }
    for (j = i + 1;j < row;j++)
    {
        count = count + 1;
        funcsubtract(j, i, count);
    }
}
```

```
    /* Print out answers depending on number of equations */

    if (n == 3)
    {
        /* 3x3 matrix */

        z = matrix[2][3];
        y = matrix[1][3] - z * matrix[1][2];
        x = matrix[0][3] - y * matrix[0][1] - z * matrix[0][2];
        printf("x = %f y= %f z = %f", x, y, z);
    }
    if (n == 4)
    {
        /* 4x4 matrix */

        d = matrix[3][4];
        c = matrix[2][4] - d * matrix[2][3];
        b = matrix[1][4] - c * matrix[1][2] - d * matrix[1][3];
        a = matrix[0][4] - b * matrix[0][1] - c * matrix[0][2] - d *
        matrix[0][3];
        printf("a = %lf b= %lf c = %lf d = %lf", a, b, c, d);
    }
    if (n == 5)
    {
        /* 5x5 matrix */

        e = matrix[4][5];
        d = matrix[3][5] - e * matrix[3][4];
        c = matrix[2][5] - d * matrix[2][3] - e * matrix[2][4];
        b = matrix[1][5] - c * matrix[1][2] - d * matrix[1][3] - e *
        matrix[1][4];
        a = matrix[0][5] - b * matrix[0][1] - c * matrix[0][2] - d *
        matrix[0][3] - e * matrix[0][4];
        printf("a = %lf b= %lf c = %lf d = %lf e = %lf", a, b, c, d, e);
    }

}
```

```
/* Function to perform division on a row */

void funcdivide(int first, int second, int count)
{

    int i, j;

    divisor = matrix[first][second];
    for (i = 0;i < col;i++)
    {
        matrix[first][i] = matrix[first][i] / divisor;
    }

    /* The next few lines are commented out. You can use them to display
    your matrix at each stage for testing */
    /* The number of terms in printf will vary with the size of the
    matrix */
    /*
        printf("augmented matrix after %d operation is\n",count);
        for(j=0;j<row;j++)
        {
            printf("%lf %lf %lf %lf\n",matrix[j][0],matrix[j]
            [1],matrix[j][2],matrix[j][3]);
        }
    */

}
/* Function to perform subtraction of one row from another */

void funcsubtract(int first, int second, int count)
{

    int i, j;

    for (i = 0;i < col;i++)
    {
        matrix[first][i] = (matrix[first][i]) - matrix[second][i];
    }

    /* The next few lines are commented out. You can use them to display
    your matrix at each stage for testing */
```

```
/* The number of terms in printf will vary with the size of the
matrix */
/*
        printf("augmented matrix after %d operation is\n",count);
        for(j=0;j<row;j++)
        {
                printf("%lf %lf %lf %lf\n",matrix[j][0],matrix[j]
                [1],matrix[j][2],matrix[j][3]);
        }
*/
}
```

Test the preceding program with the following augmented matrices. There is a 3x4, a 4x5, and a 5x6 matrix.

$$\begin{pmatrix} 2 & 3 & -4 & -4 \\ 3 & -2 & 5 & 14 \\ 4 & 5 & -6 & -4 \end{pmatrix}$$

Your answers should be a=1, b=2, and c=3.

$$\begin{pmatrix} 4 & 3 & 2 & 1 & 20 \\ 5 & 4 & -3 & -2 & -4 \\ 6 & 5 & -5 & -3 & -11 \\ 7 & 7 & -7 & -5 & -20 \end{pmatrix}$$

Your answers should be a=1, b=2, c=3, and d=4

$$\begin{pmatrix} 5 & 3 & -2 & 1 & -7 & -26 \\ 2 & -5 & 3 & -2 & 4 & 13 \\ 3 & -2 & 5 & -7 & 2 & -4 \\ 4 & 1 & 3 & -4 & -3 & -16 \\ 2 & 4 & 4 & -5 & -5 & -23 \end{pmatrix}$$

Answers are a=1, b=2, c=3, d=4, and e=5.

And then try this one.

$$\begin{pmatrix} 2 & 3 & -6 & 5 & 10 \\ 4 & 6 & -5 & -3 & -11 \\ 4 & 5 & -6 & 2 & 4 \\ 3 & 2 & 5 & -7 & -6 \end{pmatrix}$$

This augmented matrix should have produced an error. This is a problem with this technique. What happens is that while doing the operations on the rows and columns, you may get a situation where you get a zero as one of the elements. For instance, if on one of your subtraction procedures you subtract 2 from 2, you will get zero as that element. If you then go on to your divide function and try to divide by that zero, you will get an error. This is usually an INF error (as any division of a nonzero number by zero gives infinity).

We have to put in some extra code to check for this. One method to deal with it is to swop the whole row with the one below. You then get the same problem when you do the division again so you swop again. Eventually the row with the zero ends up as the last row. You can check for this as zero and not divide.

The following diagrams illustrate the procedure.

Here our equations are

$$2a + 2b + c = 9$$

$$3a + 3b - c = 6$$

$$5a - 5b + c = -2$$

So our 3x4 augmented matrix is

$$\begin{pmatrix} 2 & 2 & 1 & 9 \\ 3 & 3 & -1 & 6 \\ 5 & -5 & 1 & -2 \end{pmatrix}$$

Going through our nine stages, we first divide each row by the first element in that row, giving

$$\begin{pmatrix} 1 & 1 & 1/2 & 9/2 \\ 1 & 1 & -1/3 & 6/3 \\ 1 & -1 & \dfrac{1}{5} & -2/5 \end{pmatrix} = \begin{pmatrix} 1 & 1 & 0.5 & 4.5 \\ 1 & 1 & -0.3 & 2 \\ 1 & -1 & 0.2 & -0.4 \end{pmatrix}$$

Then we subtract the first row from the second and the third, giving

$$\begin{pmatrix} 1 & 1 & 1/2 & 9/2 \\ 0 & 0 & \left(-\dfrac{1}{3}-\dfrac{1}{2}\right) & \left(\dfrac{6}{3}-\dfrac{9}{2}\right) \\ 0 & -2 & \left(\dfrac{1}{5}-\dfrac{1}{2}\right) & \left(-\dfrac{2}{5}-\dfrac{9}{2}\right) \end{pmatrix} = \begin{pmatrix} 1 & 1 & 0.5 & 4.5 \\ 0 & 0 & -0.83 & -2.5 \\ 0 & -2 & -0.3 & -4.9 \end{pmatrix}$$

This is where we hit the snag. The next stage says divide by the second element of each row, but in the second row, this is zero – so this would crash the program. So we swop rows 2 and 3, giving

$$\begin{pmatrix} 1 & 1 & 1/2 & 9/2 \\ 0 & -2 & \left(\dfrac{1}{5}-\dfrac{1}{2}\right) & \left(-\dfrac{2}{5}-\dfrac{9}{2}\right) \\ 0 & 0 & \left(-\dfrac{1}{3}-\dfrac{1}{2}\right) & \left(\dfrac{6}{3}-\dfrac{9}{2}\right) \end{pmatrix} = \begin{pmatrix} 1 & 1 & 0.5 & 4.5 \\ 0 & -2 & -0.3 & -4.9 \\ 0 & 0 & -0.83 & -2.5 \end{pmatrix}$$

Now we have –2 as the second element of the second row so we divide by it, giving

$$\begin{pmatrix} 1 & 1 & 1/2 & 9/2 \\ 0 & 1 & \left(\dfrac{1}{5}-\dfrac{1}{2}\right)/-2 & \left(-\dfrac{2}{5}-\dfrac{9}{2}\right)/-2 \\ 0 & 0 & \left(-\dfrac{1}{3}-\dfrac{1}{2}\right) & \left(\dfrac{6}{3}-\dfrac{9}{2}\right) \end{pmatrix} = \begin{pmatrix} 1 & 1 & 0.5 & 4.5 \\ 0 & 1 & 0.15 & 2.45 \\ 0 & 0 & -0.83 & -2.5 \end{pmatrix}$$

Finally we can divide the third row by its third element, giving

$$\begin{pmatrix} 1 & 1 & 0.5 & 4.5 \\ 0 & 1 & 0.15 & 2.45 \\ 0 & 0 & 1 & 3 \end{pmatrix}$$

So we have completed the mechanism. We can rewrite our original equations using the preceding augmented matrix, giving

$$1a + 1b + 0.5c = 4.5$$

$$1b + 0.15c = 2.45$$

$$1c = 3$$

So as the third equation tells us $c = 3$, we can substitute this into the second equation, giving

$$b + 0.15*3 = 2.45$$

$$\text{or } b = 2.45 - 0.45$$

$$\text{or } b = 2$$

Now we can substitute our values of b and c into the first equation, giving

$$a + 2 + 0.5*3 = 4.5$$

$$\text{or } a = 4.5 - 2 - 1.5$$

$$\text{or } a = 1$$

So our answers are $a = 1$, $b = 2$, and $c = 3$.

The next piece of code has this check for dividing by zero. It is in the funcdivide function. It uses the array "swopmatrix" in the swopping mechanism.

```
/* augmat18 */
/* augmented matrix for 3,4 or 5 equations */
/* calls functions for row division and row subtraction */
/* catches incidences of zeros which would cause program to crash (3x3 ,4x4
and 5x5) */
```

```c
#define _CRT_SECURE_NO_WARNINGS
#include <stdio.h>
#include <math.h>

void funcdivide(int first, int second, int count);
void funcsubtract(int first, int second, int count);

double matrix[12][13];
double divisor;

int i, row, col;

main()
{
    double value, x, y, z, a, b, c, d, e;

    int i, j, k, n, count;
    /* Enter data for augmented matrix - one element at a time*/

    printf("enter row/column number (square matrix 3x3 4x4 5x5 only)");
    scanf("%d", &n);

    printf("square matrix is %d", n);

    row = n;
    col = n + 1;

    for (j = 0;j < n;j++)
    {
        printf("row %d ", j);
        for (i = 0;i < n + 1;i++)
        {
            printf("enter x\n");
            scanf("%lf", &value);
            matrix[j][i] = value;
        }
    }
    /* Print entered matrix */
```

```
    printf("augmented matrix is\n");
    for (j = 0;j < n;j++)
    {

        for (i = 0;i < n + 1;i++)
        {
            printf("matrix[%d][%d] = %lf\n", j, i, matrix[j][i]);

        }

    }


/* Calculate row-reduced form of matrix */
/* Set leading number in each row to 1 */
/* Subtract rows */

    count = 0;
    for (i = 0;i < col;i++)
    {
        for (k = i;k < row;k++)
        {
            count = count + 1;
            funcdivide(k, i, count);
        }

        for (j = i + 1;j < row;j++)
        {
            count = count + 1;
            funcsubtract(j, i, count);
        }

    }


/* Print out answers depending on number of equations */
if (n == 3)
{
    z = matrix[2][3];
    y = matrix[1][3] - z * matrix[1][2];
```

```
            x = matrix[0][3] - y * matrix[0][1] - z * matrix[0][2];
            printf("x = %f y= %f z = %f", x, y, z);
    }
    if (n == 4)
    {
            d = matrix[3][4];
            c = matrix[2][4] - d * matrix[2][3];
            b = matrix[1][4] - c * matrix[1][2] - d * matrix[1][3];
            a = matrix[0][4] - b * matrix[0][1] - c * matrix[0][2] - d *
            matrix[0][3];
            printf("a = %lf b= %lf c = %lf d = %lf", a, b, c, d);
    }
    if (n == 5)
    {
            e = matrix[4][5];
            d = matrix[3][5] - e * matrix[3][4];
            c = matrix[2][5] - d * matrix[2][3] - e * matrix[2][4];
            b = matrix[1][5] - c * matrix[1][2] - d * matrix[1][3] - e *
            matrix[1][4];
            a = matrix[0][5] - b * matrix[0][1] - c * matrix[0][2] - d *
            matrix[0][3] - e * matrix[0][4];
            printf("a = %lf b= %lf c = %lf d = %lf e = %lf", a, b, c, d, e);
    }

}


/* Function to perform division on a row */
void funcdivide(int first, int second, int count)
{

    int i, j;
    double swopmatrix[12][13];

a1:divisor = matrix[first][second];
    printf("divisor is %lf \n", divisor);
    if (divisor == 0.0 && first != row - 1)      /* check for divisor of
                                                    zero (don't try to swop if
                                                    last row) */
```

```
{

    /* zero divisor so swop this row with the one below */
    for (i = 0;i < col;i++)
    {

        swopmatrix[0][i] = matrix[first][i];
    }
    for (i = 0;i < col;i++)
    {

        matrix[first][i] = matrix[first + 1][i];
    }
    for (i = 0;i < col;i++)
    {

        matrix[first + 1][i] = swopmatrix[0][i];
    }
    /* The next few lines are commented out. You can use them to
    display your matrix at each stage for testing */
    /* The number of terms in printf will vary with the size of the
    matrix */
    /*

                printf("augmented matrix after swop  operation
                is\n");
                for(j=0;j<row;j++)
                {
                    if(row == 3)
                        printf("%lf %lf %lf %lf\n",matrix[j]
                        [0],matrix[j][1],matrix[j][2],matrix[j]
                        [3]);
                    else if(row == 4)

                        printf("%lf %lf %lf %lf %lf\n",
                        matrix[j][0],matrix[j][1],matrix[j][2],
                        matrix[j][3],matrix[j][4]);
```

```
                              else
                                  printf("%lf %lf %lf %lf %lf %lf\
                                  n",matrix[j][0],matrix[j][1],
                                  matrix[j][2],matrix[j][3],
                                  matrix[j][4],matrix[j][5]);
                    }
        */

        divisor = matrix[first][second];
        goto a1; /* check if the next element is a zero divisor */
}

if (divisor == 0.0 && first == row - 1)
{
        /* Check if last row with zero divisor */
        printf("ignore zero divisor\n");
}
else
        for (i = second;i < col;i++)
        {
                matrix[first][i] = matrix[first][i] / divisor;
        }

printf("augmented matrix after %d operation is\n", count);
for (j = 0;j < row;j++)
{
        if (row == 3)
                printf("%lf %lf %lf %lf\n", matrix[j][0], matrix[j][1],
                matrix[j][2], matrix[j][3]);
        else if (row == 4)
                printf("%lf %lf %lf %lf %lf\n", matrix[j][0], matrix[j]
                [1], matrix[j][2], matrix[j][3], matrix[j][4]);
```

```
        else
                printf("%lf %lf %lf %lf %lf %lf\n", matrix[j][0],
                matrix[j][1], matrix[j][2], matrix[j][3], matrix[j][4],
                matrix[j][5]);
    }
}

/* Function to perform subtraction of one row from another */
void funcsubtract(int first, int second, int count)
{

    int i, j;

    if (matrix[first][second] != 0.000000) /* check for swopped row */
        for (i = second;i < col;i++)
        {
                matrix[first][i] = (matrix[first][i]) - matrix[second][i];
        }

    /* The next few lines are commented out. You can use them to display
    your matrix at each stage for testing */
    /* The number of terms in printf will vary with the size of the
    matrix */
    /*
        printf("augmented matrix after %d operation is\n",count);
        for(j=0;j<row;j++)
        {
            if(row == 3)
                printf("%lf %lf %lf %lf\n",matrix[j][0],matrix[j]
                [1],matrix[j][2],matrix[j][3]);
            else if(row == 4)
                printf("%lf %lf %lf %lf %lf\n",matrix[j][0],
                matrix[j][1],matrix[j][2],matrix[j][3],matrix[j][4]);
```

```
                    else
                            printf("%lf %lf %lf %lf %lf %lf\n",matrix[j]
                            [0],matrix[j][1],matrix[j][2],matrix[j][3],
                            matrix[j][4],matrix[j][5]);
                }
        */
}
```

Test this out with the augmented matrix that failed on the previous program. You should get the answers a=1, b=2, c=3, and d=4.

Twelve Simultaneous Equations

The programs up to now have been to illustrate how the process of row reduction on the augmented matrix works. The point of this technique is to solve large numbers of simultaneous equations.

The limitations are really only set by the size of the variables that are allowed on the computer. We have *set* our limit in the programs here to solving 12 simultaneous equations. This is set by the matrix[12][13] definition in the program which limits us to 12 rows and 13 columns in our augmented matrix.

The following is the augmented matrix for the 12 simultaneous equations we want to solve.

$$
\begin{array}{cccccccccccccc}
2.6 & 3.1 & 7.4 & 0.6 & 9.3 & 4.9 & 3.4 & 8.7 & 0.2 & 3.6 & 7.7 & 3.9 & 394.125 \\
4.9 & 9.3 & 0.6 & 7.4 & 3.1 & 2.6 & 0.3 & 6.3 & 5.1 & 4.9 & 9.1 & 0.6 & 360.703 \\
8.3 & 8.8 & 5.2 & 2.7 & 0.8 & 1.3 & 8.5 & 7.6 & 6.2 & 4.1 & 0.4 & 1.2 & 324.618 \\
1.3 & 0.4 & 2.3 & 5.8 & 8.1 & 6.3 & 6.3 & 5.1 & 9.2 & 6.6 & 1.3 & 2.3 & 414.999 \\
9.7 & 6.8 & 3.9 & 0.4 & 6.7 & 4.1 & 7.1 & 6.3 & 5.5 & 4.1 & 1.7 & 3.1 & 366.599 \\
7.3 & 5.8 & 6.1 & 2.7 & 9.2 & 1.8 & 4.2 & 5.2 & 7.1 & 3.7 & 2.9 & 2.6 & 364.771 \\
9.2 & 7.3 & 9.3 & 2.4 & 3.6 & 1.2 & 2.6 & 3.7 & 6.2 & 2.7 & 3.1 & 0.2 & 273.245 \\
8.6 & 8.4 & 8.7 & 6.8 & 3.9 & 4.3 & 3.8 & 4.6 & 5.3 & 6.5 & 2.6 & 0.5 & 359.206 \\
9.4 & 9.3 & 7.1 & 7.3 & 2.4 & 3.1 & 9.2 & 8.3 & 7.1 & 6.7 & 4.1 & 1.3 & 463.157 \\
7.2 & 6.8 & 7.6 & 3.5 & 3.1 & 2.5 & 3.9 & 7.6 & 8.3 & 8.5 & 5.2 & 2.1 & 445.064 \\
6.9 & 9.9 & 8.4 & 7.7 & 4.1 & 3.8 & 8.2 & 9.7 & 6.5 & 7.7 & 1.3 & 1.8 & 465.479 \\
5.9 & 8.1 & 6.8 & 4.6 & 1.6 & 2.2 & 7.9 & 6.4 & 8.5 & 5.9 & 0.6 & 1.6 & 377.258
\end{array}
$$

The following is the program we can use to do this. We preset the matrix and copy it into the array "matrix".

```
/* augmat19C*/
/* augmented matrix for 12 equations */
/* presets the 12x13 augmented matrix */
/* calls functions for row division and row subtraction        */
/* catches incidences of zeros which would cause program to crash   */

#include <stdio.h>
#include <math.h>
#include <stdlib.h>
        void funcdivide(int first,int second,int count);
        void funcsubtract(int first,int second,int count);
        double matrix[12][13];
        double divisor;
        int i,row,col,rw,cl;

main()
{

        double solution[12];

        int i,j,k,n,count;

            double inmat[12][13]={
        {2.6,3.1,7.4,0.6,9.3,4.9,3.4,8.7,0.2,3.6,7.7,3.9,394.125},
        {4.9,9.3,0.6,7.4,3.1,2.6,0.3,6.3,5.1,4.9,9.1,0.6,360.703},
        {8.3,8.8,5.2,2.7,0.8,1.3,8.5,7.6,6.2,4.1,0.4,1.2,324.618},
        {1.3,0.4,2.3,5.8,8.1,6.3,6.3,5.1,9.2,6.6,1.3,2.3,414.999},
        {9.7,6.8,3.9,0.4,6.7,4.1,7.1,6.3,5.5,4.1,1.7,3.1,366.599},
        {7.3,5.8,6.1,2.7,9.2,1.8,4.2,5.2,7.1,3.7,2.9,2.6,364.771},
        {9.2,7.3,9.3,2.4,3.6,1.2,2.6,3.7,6.2,2.7,3.1,0.2,273.245},
        {8.6,8.4,8.7,6.8,3.9,4.3,3.8,4.6,5.3,6.5,2.6,0.5,359.206},
        {9.4,9.3,7.1,7.3,2.4,3.1,9.2,8.3,7.1,6.7,4.1,1.3,463.157},
        {7.2,6.8,7.6,3.5,3.1,2.5,3.9,7.6,8.3,8.5,5.2,2.1,445.064},
        {6.9,9.9,8.4,7.7,4.1,3.8,8.2,9.7,6.5,7.7,1.3,1.8,465.479},
        {5.9,8.1,6.8,4.6,1.6,2.2,7.9,6.4,8.5,5.9,0.6,1.6,377.258}
        };
```

```
        /* Use preset array */

        n=12;/* set matrix to 12x12 */

        /* Copy preset array to output array */

        for(i=0;i<12;i++)
        {
                for(j=0;j<13;j++)
                {
                        matrix[i][j]=inmat[i][j];
                }

        }

    /* Print preset matrix */

    printf("augmented matrix is\n");
    for(j=0;j<n;j++)
    {
        for(i=0;i<n+1;i++)
        {
                printf("matrix[%d][%d] = %lf\n",j,i,matrix[j][i]);

        }
    }

    row=n; /* row count */
    col=n+1; /* column count */

/* Call functions. One to divide a row by an element of that row */
/* Tne other to subtract one row from another */

    count=0;
    for(i=0;i<col;i++)
    {
        for(k=i;k<row;k++)
        {
                count=count+1;
                funcdivide(k,i,count);
        }
```

```
        for(j=i+1;j<row;j++)
        {
                count=count+1;
                funcsubtract(j,i,count);
        }
}

/* Calculate solutions from reduced augmented matrix */

rw=n-1;
cl=n;
solution[rw]=matrix[rw][cl];

for(i=1;i<=cl;i++)
{
        solution[rw-i]=matrix[rw-i][cl];
        j=rw;
        for(k=0;k<=i;k++)
        {
                solution[rw-i]=solution[rw-i]-matrix[rw-i][rw-i+1+k]*
                solution[rw-i+1+k];
                j=j-1;
        }
}
printf("solution\n");
for(i=0;i<=rw;i++)
{
        printf("\n x%d = %lf",i+1, solution[i]);
}
}

/* Function to perform division on a row */
void funcdivide(int first,int second,int count)
{

        int i,j;
        double swopmatrix[12][13];

        a1:divisor=matrix[first][second];
```

193

```c
if(divisor == 0.0 && first != row-1)    /* check for divisor of zero
(don't try to swop if last row) */
{
        /* zero divisor so swop this row with the one below */
        for(i=0;i<col;i++)
        {
                swopmatrix[0][i]=matrix[first][i];
        }
        for(i=0;i<col;i++)
        {
                matrix[first][i]=matrix[first+1][i];
        }
        for(i=0;i<col;i++)
        {
                matrix[first+1][i]=swopmatrix[0][i];
        }
        printf("augmented matrix after swop  operation is\n");
        for(j=0;j<row;j++)
        {
                if(row == 3)
                        printf("%lf %lf %lf %lf\n",matrix[j][0],matrix[j]
                        [1],matrix[j][2],matrix[j][3]);
                else if(row == 4)

                        printf("%lf %lf %lf %lf %lf\n",matrix[j][0],
                        matrix[j][1],matrix[j][2],matrix[j][3],matrix[j][4]);
                        else
                                printf("%lf %lf %lf %lf %lf %lf\n",matrix[j]
                                [0],matrix[j][1],matrix[j][2],matrix[j][3],
                                matrix[j][4],matrix[j][5]);
        }
    divisor=matrix[first][second];
    goto a1; /* check if the next element is a zero divisor */
    }
```

```
    if(divisor == 0.0 && first == row-1)
    {
            /* Check if last row with zero divisor */
            printf("ignore zero divisor\n");
    }else
            for(i=second;i<col;i++)
            {
                    matrix[first][i]=matrix[first][i]/divisor;
            }
```

/* The next few lines are commented out. You can use them to display your matrix at each stage for testing */
/* The number of terms in printf will vary with the size of the matrix */
/* printf("augmented matrix after %d operation is\n",count);

```
    for(j=0;j<row;j++)
    {
            printf("%lf %lf %lf %lf %lf %lf %lf\n",matrix[j][0],matrix[j]
            [1],matrix[j][2],matrix[j][3],matrix[j][4],matrix[j]
            [5],matrix[j][6]);
    }
```

*/

```
}
```

/* Function to perform subtraction of one row from another */

```
void funcsubtract(int first,int second,int count)
{

    int i;

    for(i=0;i<col;i++)
    {
            matrix[first][i]=(matrix[first][i])-matrix[second][i];
    }
```

/* The next few lines are commented out. You can use them to display your matrix at each stage for testing */
/* The number of terms in printf will vary with the size of the matrix */
/*

```
        printf("augmented matrix after %d operation is\n",count);
        for(j=0;j<row;j++)
        {
                printf("%lf %lf %lf %lf %lf %lf %lf\n",matrix[j][0],matrix[j]
                [1],matrix[j][2],matrix[j][3],matrix[j][4],matrix[j][5],
                matrix[j][6]);
        }
*/
}
```

If you run this program, you should get the result

X1 = 1.1

X2 = 2.2

X3 = 3.3

X4 = 4.4

X5 = 5.5

X6 = 6.6

X7 = 7.7

X8 = 8.8

X9 = 9.9

X10 = 10.10

X11 = 11.11

X12 = 12.12

In various applications, you can read data, like the data for the preceding augmented matrix, from a file. The next chapter is concerned with file operations and gives an example of how you could read this data in. You can also have the data in a preset array in your program.

In the last program of this chapter, the program gives you the option of entering the data manually or using the preset array.

Preset arrays like the one used in the following program are commonplace in software.

```
/* augmat25 */
/* augmented matrix for up to 12 equations */
/* calls functions for row division and row subtraction      */
/* catches incidences of zeros which would cause program to crash  */
```

```c
/* preset array for augmented matrix */
/* Can use prest array or enter data manually */

#define _CRT_SECURE_NO_WARNINGS
#include <stdio.h>
#include <math.h>
#include <stdlib.h>

struct record
{
      double matrix[12][13];
};

void funcdivide(int first, int second, int count);
void funcsubtract(int first, int second, int count);

double matrix[12][13];
double divisor;
int i, row, col, rw, cl;

main()
{

      double value;
      double solution[12];
      int j, k, /*m,*/ n, count;
      int pr;

      double inmat[12][13] = {
{2.6,3.1,7.4,0.6,9.3,4.9,3.4,8.7,0.2,3.6,7.7,3.9,394.125},
{4.9,9.3,0.6,7.4,3.1,2.6,0.3,6.3,5.1,4.9,9.1,0.6,360.703},
{8.3,8.8,5.2,2.7,0.8,1.3,8.5,7.6,6.2,4.1,0.4,1.2,324.618},
{1.3,0.4,2.3,5.8,8.1,6.3,6.3,5.1,9.2,6.6,1.3,2.3,414.999},
{9.7,6.8,3.9,0.4,6.7,4.1,7.1,6.3,5.5,4.1,1.7,3.1,366.599},
{7.3,5.8,6.1,2.7,9.2,1.8,4.2,5.2,7.1,3.7,2.9,2.6,364.771},

{9.2,7.3,9.3,2.4,3.6,1.2,2.6,3.7,6.2,2.7,3.1,0.2,273.245},
{8.6,8.4,8.7,6.8,3.9,4.3,3.8,4.6,5.3,6.5,2.6,0.5,359.206},
{9.4,9.3,7.1,7.3,2.4,3.1,9.2,8.3,7.1,6.7,4.1,1.3,463.157},
```

```
{7.2,6.8,7.6,3.5,3.1,2.5,3.9,7.6,8.3,8.5,5.2,2.1,445.064},
{6.9,9.9,8.4,7.7,4.1,3.8,8.2,9.7,6.5,7.7,1.3,1.8,465.479},
{5.9,8.1,6.8,4.6,1.6,2.2,7.9,6.4,8.5,5.9,0.6,1.6,377.258}
    };

    printf("enter 1 or 2 (1=use preset matrix, 2=enter matrix
    manually))");
    scanf("%d", &pr);
    if (pr == 2)
    {
        /* Enter data manually */

        printf("enter row/column number (square matrix only up to 12
        rows)");
        scanf("%d", &n);

        printf("square matrix is %d", n);

        row = n;
        col = n + 1;
        for (j = 0;j < n;j++)
        {
            printf("row %d ", j);
            for (i = 0;i < n + 1;i++)
            {
                printf("enter x\n");
                scanf("%lf", &value);
                matrix[j][i] = value;
            }
        }

    }
    else
    {
        /* Use preset array */

        n = 12;/* set matrix to 12x12 */

        /* Copy preset array to output array */
```

```
        for (i = 0;i < 12;i++)
        {
                for (j = 0;j < 13;j++)
                {
                        matrix[i][j] = inmat[i][j];
                }

        }

    }
    /* Print matrix  */

    for (i = 0;i < 12;i++)
    {
        for (j = 0;j < 13;j++)
        {
                printf("matrix[%d][%d] = %lf \n", i, j, matrix[i][j]);
        }

    }

    row = n; /* row count */
    col = n + 1; /* column count */

/* Call functions. One to divide a row by an element of that row */
/* Tne other to subtract one row from another */

    count = 0;
    for (i = 0;i < col;i++)
    {
        for (k = i;k < row;k++)
        {
                count = count + 1;
                funcdivide(k, i, count);
        }
```

```
            for (j = i + 1;j < row;j++)
            {
                    count = count + 1;
                    funcsubtract(j, i, count);
            }
    }

    /* Calculate solutions from reduced augmented matrix */

    rw = n - 1;
    cl = n;
    solution[rw] = matrix[rw][cl];

    for (i = 1;i <= cl;i++)
    {
        solution[rw - i] = matrix[rw - i][cl];
        j = rw;
        for (k = 0;k <= i;k++)
        {
                solution[rw - i] = solution[rw - i] - matrix[rw - i]
                [rw - i + 1 + k] * solution[rw - i + 1 + k];
                j = j - 1;
        }
    }
    printf("solution\n");
    for (i = 0;i <= rw;i++)
    {
        printf("\n x%d = %lf", i + 1, solution[i]);
    }
}

/* Function to perform division on a row */

void funcdivide(int first, int second, int count)
{

    int i, j;
    double swopmatrix[12][13];
```

```
a1:divisor = matrix[first][second];

    if (divisor == 0.0 && first != row - 1)    /* check for divisor of
    zero (don't try to swop if last row) */
    {
        /* zero divisor so swop this row with the one below */
        for (i = 0;i < col;i++)
        {

            swopmatrix[0][i] = matrix[first][i];
        }
        for (i = 0;i < col;i++)
        {

            matrix[first][i] = matrix[first + 1][i];
        }
        for (i = 0;i < col;i++)
        {

            matrix[first + 1][i] = swopmatrix[0][i];
        }

        printf("augmented matrix after swop  operation is\n");
        for (j = 0;j < row;j++)
        {
            if (row == 3)
                printf("%lf %lf %lf %lf\n", matrix[j][0],
                matrix[j][1], matrix[j][2], matrix[j][3]);
            else if (row == 4)
                printf("%lf %lf %lf %lf %lf\n", matrix[j][0],
                matrix[j][1], matrix[j][2], matrix[j][3],
                matrix[j][4]);
            else
                printf("%lf %lf %lf %lf %lf %lf\n", matrix[j]
                [0], matrix[j][1], matrix[j][2], matrix[j][3],
                matrix[j][4], matrix[j][5]);
        }
```

```
        divisor = matrix[first][second];
        goto a1; /* check if the next element is a zero divisor */
}

if (divisor == 0.0 && first == row - 1)
{
        /* Check if last row with zero divisor */
        printf("ignore zero divisor\n");
}
else
        for (i = second;i < col;i++)
        {
                matrix[first][i] = matrix[first][i] / divisor;
        }
/* The next few lines are commented out. You can use them to display
your matrix at each stage for testing */
/* The number of terms in printf will vary with the size of the
matrix */
/*
        printf("augmented matrix after %d operation is\n",count);
        for(j=0;j<row;j++)
        {
                printf("%lf %lf %lf %lf %lf %lf %lf %lf %lf %lf %lf %lf
                %lf\n",matrix[j][0],matrix[j][1],matrix[j][2],matrix[j]
                [3],matrix[j][4],matrix[j][5],matrix[j][6],matrix[j]
                [7],matrix[j][8],matrix[j][9],matrix[j][10],matrix[j]
                [11],matrix[j][12]);
        }
*/
}

/* Function to perform subtraction of one row from another */
```

```
void funcsubtract(int first, int second, int count)
{

    int i;

    for (i = 0;i < col;i++)
    {
    matrix[first][i] = (matrix[first][i]) - matrix[second][i];
    }

    /* The next few lines are commented out. You can use them to display
    your matrix at each stage for testing */
    /* The number of terms in printf will vary with the size of the
    matrix */
    /*
        printf("augmented matrix after %d operation is\n",count);
        for(j=0;j<row;j++)
        {
            printf("%lf %lf %lf %lf %lf %lf %lf %lf %lf %lf %lf %lf
            %lf\n",matrix[j][0],matrix[j][1],matrix[j][2],matrix[j]
            [3],matrix[j][4],matrix[j][5],matrix[j][6],matrix[j]
            [7],matrix[j][8],matrix[j][9],matrix[j][10],matrix[j]
            [11],matrix[j][12]);
        }
    */
}
```

EXERCISES

Use one of your augmented matrix programs to solve the following simultaneous equations.

1.

$$5.2a + 2.7b - 3.4c = -2.48$$

$$7.3a - 0.9b - 2.1c = -2.18$$

$$6.7a - 8.1b + 1.9c = -1.7$$

2.

$$2.3a - 3.1b - 4.2c = -18.15$$

$$3.2a - 1.4b + 3.2c = 11$$

$$4.9a + 2.6b - 0.2c = 10.45$$

3.

$$2.6a + 3.1b + 7.4c + 0.6d + 9.3e + 4.9f = 120.23$$

$$4.9a + 9.3b + 0.6c + 7.4d + 3.1e + 2.6f = 94.6$$

$$8.3a + 8.8b + 5.2c + 2.7d + 0.8e + 1.3f = 70.51$$

$$1.3a + 0.4b + 2.3c + 5.8d + 8.1e + 6.3f = 121.55$$

$$9.7a + 6.8b + 3.9c + 0.4d + 6.7e + 4.1f = 104.17$$

$$7.3a + 5.8b + 6.1c + 2.7d + 9.2e + 1.8f = 115.28$$

CHAPTER 9

File Access

This chapter is about moving data to and from files. The basic commands of file access are fopen (which opens a file), fclose (which closes it), fread (which reads data from a file which has been opened), and fwrite (which writes data to a file which has been opened). There are one or two other file commands which we shall meet later.

We have already come across some of our file access commands in earlier chapters. In Chapter 7 when we were looking at our radioactivity simulation program, we wrote output data to a file which could then be used to input into the Graph package as data points for a graph. In Chapter 8 we needed to input a large 12x13 array of data into the program. We did this by presetting the array in the program. What we could have done was to have written a separate program which wrote the same array to a file – then our first program could have just read in the data from that file. This idea is useful if different programs need to read the same data. So we just create the file once and then any program could read it.

We will have a look at these techniques here.

First Program to Write a File

In Chapter 7 in our radioact program, we wrote the coordinates of our graph to a file called "radioact.dat". In the program we have a pointer which points to the file we are accessing. We declare the pointer at the start of the program using the instruction FILE *fptr. The asterisk, *, signifies that the variable is a pointer. Its name is fptr and whenever we access the file in our program, we use fptr. We set up the value of the pointer using the fopen command. In the program we had

```
fptr = fopen("radioact.dat","w");
```

© Philip Joyce 2019
P. Joyce, *Numerical C*, https://doi.org/10.1007/978-1-4842-5064-8_9

This says that we want to create a file called "radioact.dat". The "w" means we want write access to the file. The fopen command returns a pointer and this is stored in fptr. The code is shown as follows.

```c
/*radioact4A */
/* radioactive decay simulation */
#define _CRT_SECURE_NO_WARNINGS
#include <stdio.h>
#include <math.h>
#include <stdlib.h>
#include <time.h>
main()
{
        int j,timelimit,nuc;

        double randnumber,timeinc,lambda,timecount,probunittime;
        FILE *fptr; /* pointer to file */
        time_t  t;
        srand((unsigned) time(&t)); /* random number generator seed */
        fptr=fopen("radioact.dat","w");

        /* Ask user to input specific data */
        /* initial number of nuclei, the value of lambda, time for experiment */

        printf("Enter initial number of nuclei : ");
        scanf("%d",&nuc);

        printf("Enter lambda : ");
        scanf("%lf",&lambda);

        printf("Enter time  : ");
        scanf("%d",&timelimit);

        /* time increment of loop */

        timeinc=0.001/lambda;

        printf("Time increment :%lf",timeinc);

        /* (delta t * lambda) */
```

```
probunittime=0.001*lambda;
timecount=0;

/* Monte Carlo loop */
while(timecount<=timelimit)
{
        fprintf(fptr,"%lf %d\n",timecount,nuc);/* write two items to
        file */

        timecount=timecount+timeinc;

        for(j=0;j<=nuc;j++)
        {
                randnumber=rand()%1000;
                randnumber=randnumber/1000;

        /* Monte Carlo method checks random number less than (delta t
        * lambda) */

                if(randnumber<=probunittime)
                        nuc=nuc-1;/* If less, then prob. that nucleus has
                        decayed */

                if(nuc<=0)
                        goto nuclimitreached;

        }
}
nuclimitreached:        fclose(fptr); /* nuclei limit or time limit reached */

}
```

In this program we are using the command fprintf to write to the file. This is almost the same in form as printf except that instead of writing the data to the screen, it writes it to a file. So here we had

```
fprintf(fptr,"%f %d\n",timecount,nuc);
```

In appearance this is similar to printf except that we write the data to the file pointed to by fptr. We are writing the variables timecount and nuc to the file. When we have written all of our data to the file, we call

```
fclose(fptr);
```

This closes the file pointed to by fptr.

We did a similar thing in our randwalk program. Here we open a file called "randwalk.dat" and we supply the parameter "w" to the fopen call to say that we want to write to the file. Here our pointer to the file is called "output." We then use this name in our calls to fprintf to write the data and in the fclose to close the file.

The random walk program is shown as follows.

```c
/* randwalk6 */
/* simple random walk simulation in 1 dimension */
#define _CRT_SECURE_NO_WARNINGS
#include <stdio.h>
#include <math.h>
#include <stdlib.h>
#include <time.h>

FILE *output;
time_t  t;

main()
{
     int i;
     double xrand;

     double x,randwalkarr[20001];

     output= fopen ("randwalk6.dat", "w");        /* external file name */

     for (i=0; i<=20000; i++)
          randwalkarr [i]=0.0;                     /* clear array */

     srand((unsigned) time(&t));                   /* set the number generator */

        x=0.0;
```

```
    for (i=1;i<=20000; i++)
      {
        /* generate x random number */
        xrand=rand()%1000;
        xrand=xrand/1000;
        if(xrand<0.5)
             x=x+1.0;
        else
             x=x-1.0;

         randwalkarr[i] = sqrt(x*x);/* store randwalkarr to total */

    }
    /* Write values to file */
    for (i=0; i<=200; i++)
    {

         fprintf(output,"%d %lf\n", i, randwalkarr[i*100]);
    }

    fclose (output);
}
```

In this case our fprintf writes the values contained in the variables i and randwalkarr[i*100].

You can run your radioact program and randwalk program, and then after each has run, you can inspect the files that each program has produced. In each case you can look at the data collected. In the relevant chapter, you can see the graph that this data produced by importing it into the Graph package.

Writing a Large Data File

Our next program writes the 12x13 array, we used in our matrix chapter, to a file. The array is defined in the program in the same way as it is in our program in Chapter 8. We copy this to our output array. In this program we use a structure definition. The use of this type of definition will be made clearer later in this chapter. It is basically used if we want to write the same type of data to a file many times, for instance, if we had a file containing names, ages, and examination results of people in a college. We would have a structure for each person, and the structure would contain the person's age, their name, and their examination results. The age and examination results could use an int in the structure, and their name could use a char array as shown here.

```
struct Examdata {
int age;
char name[15];
int examscore;
};
```

We would set up one of these for each student to be written to our file.

In the case of our 12x13 array, we just have the array in the structure. We reference this by data_record.matrix. In this program our output file is called "testaug.bin" and we want the file to hold binary type data so we have "wb" for write binary in the fopen command. The code is shown as follows.

```
/* filewriteE */
#define _CRT_SECURE_NO_WARNINGS
#include<stdio.h>
#include <stdlib.h>

int main()
{

    struct record
    {
        double matrix[12][13];
    };
    int /*counter,*/ i, j;
    FILE *ptr;
    struct record data_record;
    size_t r1;
    double inmat[12][13] = {
{2.6,3.1,7.4,0.6,9.3,4.9,3.4,8.7,0.2,3.6,7.7,3.9,394.125},
{4.9,9.3,0.6,7.4,3.1,2.6,0.3,6.3,5.1,4.9,9.1,0.6,360.703},
{8.3,8.8,5.2,2.7,0.8,1.3,8.5,7.6,6.2,4.1,0.4,1.2,324.618},
{1.3,0.4,2.3,5.8,8.1,6.3,6.3,5.1,9.2,6.6,1.3,2.3,414.999},
```

```
{9.7,6.8,3.9,0.4,6.7,4.1,7.1,6.3,5.5,4.1,1.7,3.1,366.599},
{7.3,5.8,6.1,2.7,9.2,1.8,4.2,5.2,7.1,3.7,2.9,2.6,364.771},
{9.2,7.3,9.3,2.4,3.6,1.2,2.6,3.7,6.2,2.7,3.1,0.2,273.245},
{8.6,8.4,8.7,6.8,3.9,4.3,3.8,4.6,5.3,6.5,2.6,0.5,359.206},
{9.4,9.3,7.1,7.3,2.4,3.1,9.2,8.3,7.1,6.7,4.1,1.3,463.157},
{7.2,6.8,7.6,3.5,3.1,2.5,3.9,7.6,8.3,8.5,5.2,2.1,445.064},
{6.9,9.9,8.4,7.7,4.1,3.8,8.2,9.7,6.5,7.7,1.3,1.8,465.479},
{5.9,8.1,6.8,4.6,1.6,2.2,7.9,6.4,8.5,5.9,0.6,1.6,377.258}
    };

    /* Copy preset array to output array */

    for (i = 0;i < 12;i++)
    {
        for (j = 0;j < 13;j++)
        {
            data_record.matrix[i][j] = inmat[i][j];
        }

    }

    /* Open output file (write/binary) */

    ptr = fopen("testaug.bin", "wb");
    if (!ptr)
    {
        printf("Can not open file");
        return 1;
    }

    /* Write output matrix to output file */

    r1 = fwrite(data_record.matrix, sizeof(data_record.matrix), 1, ptr);
    printf("wrote %d elements \n", r1);
    printf("size of data_record.matrix is %d \n", sizeof(data_record.
    matrix));
    /* Print matrix written to file */
```

```
for (i = 0;i < 12;i++)
{
        for (j = 0;j < 13;j++)
        {
                data_record.matrix[i][j] = inmat[i][j];
                printf("data_record.matrix[%d][%d] = %lf \n", i, j,
                data_record.matrix[i][j]);
        }

}

fclose(ptr);
return 0;
}
```

Here the file pointer is called ptr. If there are any problems calling the fopen command, it will return an error code to fptr. The command if(!ptr) checks for this, and if it gets it, it outputs an appropriate error message and closes the program. Data is written to the file in this case using the fwrite command. The fwrite command has four parameters.

```
fwrite(data_record.matrix, sizeof(data_record.matrix),1,ptr);
```

Here the first is data_record.matrix which is the structure containing the data to be written. The second parameter is the size of this structure. The third parameter is 1, meaning that we want to write one structure. The fourth parameter is the file pointer to the file we want to write to.

After the data is written, we call fclose to close the file.

We can write a program which just reads the file and writes all of the data in it to the screen as a test. This program is similar to the write program except that when we open it we use "rb" in our fopen command which says that the file is to be read and in binary. To read the data, we use fread which has the same parameters as our fwrite in our write program, but it reads the structure from the file into the input array. We reference this by data_record.matrix. The code is as follows.

```
/* filereadE */
#define _CRT_SECURE_NO_WARNINGS
#include<stdio.h>
```

```c
#include <stdlib.h>

struct record
{
    double matrix[12][13];
};

int main()
{
    int counter, i;
    FILE *ptr;
    struct record data_record;
    size_t r1;

    /* Open input file (read/binary) */

    ptr = fopen("testaug.bin", "rb");
    if (!ptr)
    {
    printf("Can not open file");
        return 1;
    }

    /* Read input matrix from input file */

    r1 = fread(data_record.matrix, sizeof(data_record.matrix), 1, ptr);
    printf("read %d elements \n", r1);
    printf("size of struct record is %d \n", sizeof(struct record));

    /* Print matrix read from file */

    for (counter = 0; counter < 12; counter++)
    {
        for (i = 0; i < 13; i++)
        {
            printf("matrix[%d][%d] = %lf \n", counter, i, data_
            record.matrix[counter][i]);
        }

    }
```

```
        fclose(ptr);
        return 0;
}
```

We print out what we have read to check that both the write program and the read program are working.

At the end we close the file.

Medical Records File

Our next program in this chapter shows how we can write a structure containing different types of data to a file. In this case the data is medical data about different people possibly registered with the same family doctor. The data contains their patient identifier, their name, and their blood pressure. The structure is shown as follows.

```
struct Patient {
        int PatientID;
        char name[13];
        int BloodPressure;
};
```

There is one of these structures for each patient. The first program creates a file containing this data. The structure data for each patient is set at the beginning of the program.

The code is shown as follows.

```
/* filewrite */
/* reads from file */
/* prints out the records sequentially */
/* Finds specific records and prints them */

#define _CRT_SECURE_NO_WARNINGS
#include<stdio.h>

struct Patient {
        int PatientID;
        char name[13];
        int BloodPressure;
};
```

```
int main()
{
     int i, numread;
     FILE *fp;
     struct Patient s1;
     struct Patient s2;
/* Preset the data for each patient */

     struct Patient s10 = { 10,"Brown       ",50 };
     struct Patient s11 = { 11,"Jones       ",51 };
     struct Patient s12 = { 12,"White       ",52 };
     struct Patient s13 = { 13,"Green       ",53 };
     struct Patient s14 = { 14,"Smith       ",54 };
     struct Patient s15 = { 15,"Black       ",55 };
     struct Patient s16 = { 16,"Allen       ",56 };
     struct Patient s17 = { 17,"Stone       ",57 };
     struct Patient s18 = { 18,"Evans       ",58 };
     struct Patient s19 = { 19,"Royle       ",59 };
     struct Patient s20 = { 20,"Stone       ",60 };
     struct Patient s21 = { 21,"Weeks       ",61 };
     struct Patient s22 = { 22,"Owens       ",62 };
     struct Patient s23 = { 23,"Power       ",63 };
     struct Patient s24 = { 24,"Bloom       ",63 };

     struct Patient s28 = { 28,"Haver       ",68 };
     struct Patient s29 = { 29,"James       ",69 };

     /* Open the Patients file */
     fp = fopen("patients.bin", "w");

     /* Write details of each patient to file*/
     /* From the structures defined above */

     fwrite(&s10, sizeof(s1), 1, fp);
     fwrite(&s11, sizeof(s1), 1, fp);
     fwrite(&s12, sizeof(s1), 1, fp);
     fwrite(&s13, sizeof(s1), 1, fp);
     fwrite(&s14, sizeof(s1), 1, fp);
```

215

```
        fwrite(&s15, sizeof(s1), 1, fp);
        fwrite(&s16, sizeof(s1), 1, fp);
        fwrite(&s17, sizeof(s1), 1, fp);
        fwrite(&s18, sizeof(s1), 1, fp);
        fwrite(&s19, sizeof(s1), 1, fp);
        fwrite(&s20, sizeof(s1), 1, fp);
        fwrite(&s21, sizeof(s1), 1, fp);
        fwrite(&s22, sizeof(s1), 1, fp);
        fwrite(&s23, sizeof(s1), 1, fp);
        fwrite(&s24, sizeof(s1), 1, fp);

        fwrite(&s28, sizeof(s1), 1, fp);
        fwrite(&s29, sizeof(s1), 1, fp);

        /* Close the file */

        fclose(fp);

        /* Reopen the file */

        fopen("patients.bin", "r");

        /* Read and print out all of the records on the file */

        for (i = 0;i < 17;i++)
        {
                numread = fread(&s2, sizeof(s2), 1, fp);/* read into
                structure s2 */

                if (numread == 1)
                {
                        /*printf( "Number of items read = %d ", numread );*/
/* reference elements of structure by s2.PatientID etc */
                        printf("\nPatientID : %d", s2.PatientID);
                        printf("\nName : %s", s2.name);
                        printf("\nBloodPressure : %d", s2.BloodPressure);
                }
```

```
        else {
                /* If an error occurred on read then print out message */
                if (feof(fp))
                        printf("Error reading patients.bin : unexpected
                        end of file fp is %p\n", fp);

                else if (ferror(fp))
                {
                        perror("Error reading patients.bin");
                }
        }
    }
    /* Close the file */
    fclose(fp);

}
```

We start by opening the file. The instruction is

```
fp = fopen("patients.bin", "w");
```

where patients.bin is the file name and fp is the file pointer.

We write to the file using several fwrite calls.

We close the file and then reopen it in order to check what we have written. In our read we have

```
numread=fread(&s2, sizeof(s2), 1, fp);
```

where numread is the number of structures read. We are expecting one structure to have been read as shown by the third parameter in our fread. If it is 1, then we print the record. If it is not 1, we check the error. By calling the command feof(fp), we can check if we have had an unexpected end of file. If so then we print out an appropriate message.

Finally we close the file.

Our next program reads and displays the data from the file. Again, we open the file, this time as read only ("r" in the open call).

The following code shows this.

```c
/* filereadCh */
/* reads from file */
/* reads and prints sequentially */
/* reads and prints specific records */

#define _CRT_SECURE_NO_WARNINGS
#include<stdio.h>

struct Patient {
    int PatientID;
    char name[13];
    int BloodPressure;
};

int main()
{
    FILE *fp;

    struct Patient s2;

    int numread, i;
    /* Open patients file */

    fp = fopen("patients.bin", "r");
    for (i = 0;i < 17;i++)
    {
        /* Read each patient data from file sequentially */
        fread(&s2, sizeof(s2), 1, fp);
        /* Print patient ID, name and Blood Pressure for each patient */

        printf("\nPatientID : %d", s2.PatientID);
        printf("\n Name : %s", s2.name);
        printf("\nBloodPressure : %d", s2.BloodPressure);

    }

    fclose(fp);
```

```
/* Re-open the patients file */
fp = fopen("patients.bin", "r");
for (i = 0;i < 17;i++)
{
        /* Search the file for patient with ID of 23 */

        fread(&s2, sizeof(s2), 1, fp);
        if (s2.PatientID == 23)
        {
                /* Found the patient. Print their name */
                printf("\nName : %s", s2.name);
                break;
        }
}
/* Go back to the beginning of the file */

fseek(fp, sizeof(s2), SEEK_END);
rewind(fp);
/* Find all patients with Blood Pressure reading above 63 */

for (i = 0;i < 17;i++)
{
fread(&s2, sizeof(s2), 1, fp);
        if (s2.BloodPressure > 63)
        {
                /* Print out name of each patient with Blood pressure
                above 63 */
                printf("\nName : %s", s2.name);

        }
}
/* Go back to the beginning of the file */

rewind(fp);

/* Read and print out the first 3 patients in the file */
```

```
    numread = fread(&s2, sizeof(s2), 1, fp);
    if (numread == 1)
    {
            printf("\nPatientID : %d", s2.PatientID);
            printf("\nName : %s", s2.name);
            printf("\nBloodPressure : %d", s2.BloodPressure);
}

    numread = fread(&s2, sizeof(s2), 1, fp);
    if (numread == 1)
    {

            printf("\nPatientID : %d", s2.PatientID);
            printf("\nName : %s", s2.name);
            printf("\nBloodPressure : %d", s2.BloodPressure);
    }
    numread = fread(&s2, sizeof(s2), 1, fp);
    if (numread == 1)
    {
            printf("\nPatientID : %d", s2.PatientID);
            printf("\nName : %s", s2.name);
            printf("\nBloodPressure : %d", s2.BloodPressure) ;
    }
    /* Close the file */

    fclose(fp);


}
```

We specify in the fread that we want to read the data into the structure in our program. Here the structure is s2, and at the top of the program, we have our structure definition as for the filewrite program. In our definition of s2, we identify it as type "structure Patient". This defines the type in the same way as int defines the type for our numread as in definitions at the top of the program.

We close the file and then reopen it to illustrate what we can do with file access operations.

Rather than closing our file and reopening it, we can call rewind which sets the file back to the beginning.

Firstly, we want to find the patient who has the PatientID of 23. We set up a forloop to read each structure in turn and check if its PatientID is 23. If it is, the program prints out the patient's name and then uses "break" to come out of the loop. We rewind the file back to the start again. This time we want to find all of the patients whose blood pressure is above 63. We, again, set up a forloop to look through each structure on the file. If the blood pressure is over 63, we print out the patient's name. This time we don't break from the forloop because there may be more than one patient with blood pressure over 63.

Lastly we rewind again and just print out the first three patients in the file and show the use of the variable numread.

Our last program shows the use of the command fseek. This command enables you to access different points within the file directly. The code is similar to the previous program, but it illustrates the usefulness of fseek.

The code is as follows.

```
/* fileseek6ra */
/* reads from file */
/* reads and prints sequentially */
/* reads and prints specific records */
/* Only does seek when finding a record (not going back to start) */
#define _CRT_SECURE_NO_WARNINGS
#include<stdio.h>

struct Patient {
        int PatientID;
        char name[13];
        int BloodPressure;
};

int main()
{
        FILE *fp;

        /*FILE *fpout;*/
        struct Patient s2;
        struct Patient s1 = { 68,"Warne        ",95 };
```

```
int numread, i;
int posn;
long int minusone = -1;
/* Open patients file */

fp = fopen("patients.bin", "r+");
for (i = 0;i < 17;i++)
{
        /* Read each patient data from file sequentially */
        fread(&s2, sizeof(s2), 1, fp);
        /* Print patient ID, name and Blood Pressure for each patient */

        printf("\nPatientID : %d", s2.PatientID);
        printf("\n Name : %s", s2.name);
        printf("\nBloodPressure : %d", s2.BloodPressure);
        posn = ftell(fp);/* Find current file position */
        printf("\n file posn is : %d", posn);/* Print current file
        position */
}

fclose(fp);

/* Re-open the patients file */

fp = fopen("patients.bin", "r+");
for (i = 0;i < 17;i++)
{
        /* Search the file for patient with ID of 23 */

        fread(&s2, sizeof(s2), 1, fp);
        if (s2.PatientID == 23)
        {
                /* Found the patient. Print their name */
                printf("\nName : %s", s2.name);
                break;
        }
}
/* Go back to the beginning of the file */
```

```
rewind(fp);
/* Find all patients with Blood Pressure reading above 63 */
for (i = 0;i < 17;i++)
{
        fread(&s2, sizeof(s2), 1, fp);
        if (s2.BloodPressure > 63)
        {
                /* Print out name of each patient with Blood pressure
                above 63 */
                printf("\nName : %s", s2.name);

        }
}
/* Go back to the beginning of the file */

rewind(fp);

/* Read and print out the first 3 patients in the file */

numread = fread(&s2, sizeof(s2), 1, fp);
if (numread == 1)
{
printf("\nPatientID : %d", s2.PatientID);
        printf("\nName : %s", s2.name);
        printf("\nBloodPressure : %d", s2.BloodPressure);

}
numread = fread(&s2, sizeof(s2), 1, fp);
if (numread == 1)
{
        printf("\nPatientID : %d", s2.PatientID);
        printf("\nName : %s", s2.name);
        printf("\nBloodPressure : %d", s2.BloodPressure);
}
```

```c
numread = fread(&s2, sizeof(s2), 1, fp);
if (numread == 1)
{

        printf("\nPatientID : %d", s2.PatientID);
        printf("\nName : %s", s2.name);
        printf("\nBloodPressure : %d", s2.BloodPressure);

}

/* Demonstrate use of fseek to move current position in the file */
/* Then overwrite the current structure with a new one */

posn = ftell(fp);/* Find current file position */
printf("\n file posn is : %d", posn);/* Print current file position */

/* File pointer is now pointing to the 4th (the next) record in the
file */

fseek(fp, minusone*sizeof(s2), SEEK_CUR);/* set it back to point at
the third */

posn = ftell(fp);/* Find current file position */
printf("\n file posn is : %d", posn);/* Print current file position */

fwrite(&s1, sizeof(s1), 1, fp);/* overwrites what was in that
position (3rd) */
fclose(fp);
fp = fopen("patients.bin", "r");
for (i = 0;i < 18;i++)
{
        /* Read each patient data from file sequentially */
        fread(&s2, sizeof(s2), 1, fp);
        /* Print patient ID, name and Blood Pressure for each patient */

        printf("\nPatientID : %d", s2.PatientID);
        printf("\n Name : %s", s2.name);
        printf("\nBloodPressure : %d", s2.BloodPressure);

}
```

```
/* Close the file */

fclose(fp);
```

}

In this program we start by reading and writing to the screen the whole file. We will need this to test what we are going to do next has worked. We now move part way down the last program where we read the first three structures from the file.

The position of the file is now at the start of the fourth structure. We use the command ftell to confirm this. ftell just returns the file's current position. We then print this out. We then call the fseek command as follows.

```
fseek(fp, -sizeof(s2), SEEK_CUR);
```

What this does is it takes the current file position (SEEK_CUR) and goes backward by one structure. This is done by the -sizeof(s2) parameters in fseek. The minus sign before sizeof means go backward. We could go backward by two structures by calling fseek with -2*sizeof(s2). Or we could go forward by three structures by calling fseek with 3*sizeof(s2). Notice that we have not got the minus sign before this parameter. In this case it will move the file position three structures forward. Notice that in the code for the fseek, rather than specify -sizeof, we use the predefined long int minusone which we preset to –1. We then say minusone*sizeof(s2).

In our program we have gone one structure back. Now we can write a new structure to the file which overwrites the existing structure at that position. Finally we can print out the file to show the effect of our overwrite. We can compare the file now to what it was when we started the program before we did the overwrite.

We can extend our file to contain more information about the patients so that we can look for trends in illnesses or the possible link between two illnesses. Our extended structure is shown as follows.

```
struct Patient {
        int PatientID;
        char name[13];
        int BloodPressure;
        char allergies;
        char leukaemia;
        char anaemia;
```

```
        char asthma;
        char epilepsy;
        char famepil;
};
```

The information for the extra elements of our structure is simply y or n to say whether the patient has ever suffered from any of the illnesses. The last element "famepil" says that if you have never suffered from epilepsy but somebody in your family has.

The program for this is as follows.

```
/* filewritepatients */
/* reads from file */
/* prints out the records sequentially */

#define _CRT_SECURE_NO_WARNINGS
#include<stdio.h>

struct Patient {
    int PatientID;
    char name[13];
    int BloodPressure;
    char allergies;
    char leukaemia;
    char anaemia;
    char asthma;
    char epilepsy;
    char famepil;
};

int main()
{
    int i, numread;
    FILE *fp;
    struct Patient s1;
    struct Patient s2;

/* Preset data for each patient */
```

```
struct Patient s10 = { 10,"Brown      ",50,'y','n','n','y','n','y' };
struct Patient s11 = { 11,"Jones      ",51,'y','y','n','y','y','n' };
struct Patient s12 = { 12,"White      ",52,'y','n','y','y','n','n' };
struct Patient s13 = { 13,"Green      ",53,'y','n','y','y','n','n' };
struct Patient s14 = { 14,"Smith      ",54,'y','y','n','y','y','n' };
struct Patient s15 = { 15,"Black      ",55,'y','n','n','y','n','n' };
struct Patient s16 = { 16,"Allen      ",56,'y','n','n','y','n','y' };
struct Patient s17 = { 17,"Stone      ",57,'y','n','n','y','y','n' };
struct Patient s18 = { 18,"Evans      ",58,'y','y','n','y','n','n' };
struct Patient s19 = { 19,"Royle      ",59,'y','n','y','y','n','n' };
struct Patient s20 = { 20,"Stone      ",60,'y','y','n','y','n','n' };
struct Patient s21 = { 21,"Weeks      ",61,'y','n','n','y','y','y' };
struct Patient s22 = { 22,"Owens      ",62,'y','n','n','y','y','y' };
struct Patient s23 = { 23,"Power      ",63,'y','n','n','y','n','n' };
struct Patient s24 = { 24,"Bloom      ",63,'y','n','y','y','n','n' };
struct Patient s28 = { 28,"Haver      ",68,'y','y','n','y','n','n' };
struct Patient s29 = { 29,"James      ",69,'y','y','n','y','n','n' };

/* Open the Patients file */

fp = fopen("patientex.bin", "w");

/* Write details of each patient to file*/
/* From the structures defined above */

fwrite(&s10, sizeof(s1), 1, fp);
fwrite(&s11, sizeof(s1), 1, fp);
fwrite(&s12, sizeof(s1), 1, fp);
fwrite(&s13, sizeof(s1), 1, fp);
fwrite(&s14, sizeof(s1), 1, fp);
fwrite(&s15, sizeof(s1), 1, fp);
fwrite(&s16, sizeof(s1), 1, fp) ;
fwrite(&s17, sizeof(s1), 1, fp);
fwrite(&s18, sizeof(s1), 1, fp);
fwrite(&s19, sizeof(s1), 1, fp);
fwrite(&s20, sizeof(s1), 1, fp);
fwrite(&s21, sizeof(s1), 1, fp);
```

```c
    fwrite(&s22, sizeof(s1), 1, fp);
    fwrite(&s23, sizeof(s1), 1, fp);
    fwrite(&s24, sizeof(s1), 1, fp);

    fwrite(&s28, sizeof(s1), 1, fp);
    fwrite(&s29, sizeof(s1), 1, fp);

    /* Close the file */

    fclose(fp);

    /* Reopen the file */

    fopen("patientex.bin", "r");

    /* Read and print out all of the records on the file */

    for (i = 0;i < 17;i++)
    {

        numread = fread(&s2, sizeof(s2), 1, fp);

        if (numread == 1)
        {
            /*printf( "Number of items read = %d ", numread );*/

            printf("\nPatientID : %d", s2.PatientID) ;
            printf("\nName : %s", s2.name);
            printf("\nBloodPressure : %d", s2.BloodPressure);
            printf("\nAllergies %c leukaemia %c anaemia %c",
            s2.allergies, s2.leukaemia, s2.anaemia);
            printf("\nAsthma %c epilepsy %c famely epilepsy %c",
            s2.asthma, s2.epilepsy, s2.famepil);
        }

        else {
            /* If an error occurred on read then print out message */

            if (feof(fp))

                printf("Error reading patients.bin : unexpected
                end of file fp is %p\n", fp);
```

```
            else if (ferror(fp))
            {
                    perror("Error reading patients.bin");
            }
        }

    }
    /* Close the file */

    fclose(fp);

}
```

We shall look at interrogating this file in our Exercises section of this chapter.

Company Records File

Our next example of file use looks at a file containing data about a number of companies. The data here is just an example of the type of information that you can store about companies and then how you can use the information to predict trends in business practice. The program shown here creates the file of data. The file is called Companyex. bin. The structure used for each company's entry in the file is shown as follows.

```
struct Company {
      int CompanyID;
      char companyname[13];
      float salesprofitpct;/* profit as a % of sales */
      float totalctrypop;/* total populations countries for sales (in
      millions) */
      float advertpct;/* Advertising as a % of sales */
      float salprofpct;/* Total salaries as a % of profit */
      float mwpct;/* Women as a % of total workers */
      float alienwpct;/* Foreign workers as a % of total */

};
```

The first element of the structure is just an ID that we can use to identify the company. We then have the company name. The other elements are the details of interest. We store data for each company about the percentage that you get if you divide your profit by the value of your sales, the total populations of all of the countries that you sell to, the percentage you get if you divide your cost in advertising by your sales, the percentage you get if you divide your total salaries bill by your sales, the percentage you get if you divide the number of women you employ by the total workforce, and finally the percentage you get if you divide the number of foreign workers by nonforeign.

Structures are set up for a number of companies and then written to the file. The program is shown as follows.

```c
/* filewriteex3 */
/* Creates Company file */
/* prints out the records sequentially */

#define _CRT_SECURE_NO_WARNINGS
#include<stdio.h>

struct Company {
    int CompanyID;
    char companyname[13];
    double salesprofitpct;/* profit as a % of sales */
    double totalctrypop;/* total populations countries for sales
    (in millions) */
    double advertpct;/* Advertising as a % of sales */
    double salprofpct;/* Total salaries as a % of profit */
    double mwpct;/* Women as a % of total workers */
    double alienwpct;/* Foreign workers as a % of total */

};

int main()
{
    int i, numread;
    FILE *fp;
    struct Company s1;
    struct Company s2;

    /* Preset a structure for each company to be written to the output file */
```

```
struct Company s10 = { 10,"Brown Co     ",20.2,402,0.3,45.5,43.2,2.7 };
struct Company s11 = { 11,"CompuFix     ",1.3,354,2.6,60.3,27.5,1.6 };
struct Company s12 = { 12,"Wall's       ",12.6,766,5.8,14.7,54.6,3.8 };
struct Company s13 = { 13,"Goldman Inc ",29.5,876,12.6,21.6,43.9,9.3 };
struct Company s14 = { 14,"Stocks LLC  ",0.7,1252,8.2,18.4,38.4,3.8 };
struct Company s15 = { 15,"Black & Blue",1.4,984,5.8,12.7,27.9,10.6 };
struct Company s16 = { 16,"Allenby      ",52.8,1325,32.9,14.3,47.2,3.9 };
struct Company s17 = { 17,"StoneWorks  ",16.3,1548,4.6,28.9,51.3,4.1 };
struct Company s18 = { 18,"Evans LLC   ",51.0,1006,19.6,51.7,43.7,11.7 };
struct Company s19 = { 19,"Royle & Co  ",19.6,983,14.3,26.2,48.1,12.3 };
struct Company s20 = { 20,"Stone Inc   ",24.8,1030,8.5,13.5,34.6,5.6 };
struct Company s21 = { 21,"WeeksAway    ",16.9,547,0.9,12.9,43.9,2.9 };
struct Company s22 = { 22,"Owens Co     ",45.7,792,2.7,31.6,33.6,1.7 };
struct Company s23 = { 23,"PowerTools  ",32.6,1563,17.5,29.3,51.8,13.3 };
struct Company s24 = { 24,"Bloom        ",27.2,1869,23.9,18.7,40.4,9.6 };

struct Company s28 = { 28,"HaverGoodOne",33.8,489,3.6,12.7,43.8,3.7 };
struct Company s29 = { 29,"James & Co  ",15.5,639,17.4,15.9,36.5,4.5 };
/* Open the Companys file */

fp = fopen("Companyex.bin", "w");

/* Write details of each Company to file*/
/* From the structures defined above */

fwrite(&s10, sizeof(s1), 1, fp) ;
fwrite(&s11, sizeof(s1), 1, fp);
fwrite(&s12, sizeof(s1), 1, fp);
fwrite(&s13, sizeof(s1), 1, fp);
fwrite(&s14, sizeof(s1), 1, fp);
fwrite(&s15, sizeof(s1), 1, fp);
fwrite(&s16, sizeof(s1), 1, fp);
fwrite(&s17, sizeof(s1), 1, fp);
fwrite(&s18, sizeof(s1), 1, fp);
fwrite(&s19, sizeof(s1), 1, fp);
fwrite(&s20, sizeof(s1), 1, fp);
fwrite(&s21, sizeof(s1), 1, fp);
fwrite(&s22, sizeof(s1), 1, fp);
```

```
fwrite(&s23, sizeof(s1), 1, fp);
fwrite(&s24, sizeof(s1), 1, fp);

fwrite(&s28, sizeof(s1), 1, fp);
fwrite(&s29, sizeof(s1), 1, fp);

/* Close the file */

fclose(fp);

/* Reopen the file */

fopen("Companyex.bin", "r");

/* Read and print out all of the records on the file */

for (i = 0;i < 17;i++)
{
        numread = fread(&s2, sizeof(s2), 1, fp);

        if (numread == 1)
        {
                printf("\nCompanyID : %d", s2.CompanyID) ;
                printf("\ncompanyname : %s", s2.companyname);
                printf("\nprofit as a percentage of sales : %lf",
                s2.salesprofitpct);
                printf("\ntotal populations countries for sales (in
                millions) %lf ", s2.totalctrypop);
                printf("\nAdvertising as a percentage of sales %lf ",
                s2.advertpct);
                printf("\nTotal salaries as a percentage of profit %lf ",
                s2.salprofpct);
                printf("\nWomen as a percentage of total workers %lf ",
                s2.mwpct);
                printf("\nForeign workers as a percentage of total %lf ",
                s2.alienwpct);

        }
```

```
        else {
                /* If an error occurred on read then print out message */

                if (feof(fp))

                        printf("Error reading Companys.bin : unexpected
                        end of file fp is %p\n", fp);

                else if (ferror(fp))
                {
                        perror("Error reading Companys.bin");
                }
        }

    }
    /* Close the file */

    fclose(fp);

}
```

After all of the structures have been written to the file, the program closes the file, then reopens it and reads and prints all of the data from the file.

We shall look at interrogating this file in our Exercises section of this chapter.

EXERCISES

1. For our extended Patients File, we want to write a program to collect important information. So we can extend our program which reads the original smaller file. Our new file should start by reading all of the structures in the file and printing out each element or each patient. We then close the file and reopen it. We now want to read each patient's data and print out any which have both asthma and epilepsy. We can print out their name. We also want to keep a count of how many people are in this set of the link between asthma and epilepsy and work out what this number is as a percentage of the total number of patients in the file.

2. We now want to do a similar thing as question 1 but for our Company file. We firstly extend our program for reading the file in this chapter so that we can read the extended file. We, again, start by reading all of the data in the file and printing it out. We then close the file and reopen it. Now we go through each company on the file and test for the companies which have a women to men percentage of over 40 and a sales profit percentage over 40. When we find these, we print them out. We also want to find what this number is as a percentage of the total number of companies and then print out this number.

CHAPTER 10

Differential Equations

The solution of differential equations is one of the most important and challenging areas of mathematics. Differential equations arise naturally in many areas of medicine, economics, physics, and engineering. They are normally introduced in schools and solved by integral calculus. This is an excellent introduction to the subject but, as we have seen in other areas of this book, in real life the analytical methods cannot be used and computational techniques are called upon.

A differential equation is formed when a normal algebraic function is differentiated. This is a technique which is used to find the rate of change of one variable with another. In schools we use x and y as the variables, but in real life they could be the rate of change of cancer growth with blood iron level or the rate of growth of a company's profits with its size. We can look at a simple algebraic function.

$$y = 2x^3 - 5x^2$$

If we differentiate this, we get

$$dy/dx = 6x^2 - 10x$$

We can write dy/dx as $f'(x)$.

So $f'(x) = 6x^2 - 10x$ is our differential equation.

This can be solved easily using calculus. However, differential equations are normally much more complicated. If we differentiate our differential equation earlier again, we would get

$$f''(x) = 12x - 10$$

Again, this is a simple equation to solve. We could differentiate this again to give us $f'''(x)$. Look at the next differential equation.

$$f'(x) = (4x - 6x^2) / \exp(3x)$$

© Philip Joyce 2019
P. Joyce, *Numerical C*, https://doi.org/10.1007/978-1-4842-5064-8_10

This is much more difficult to solve using calculus, but we will solve this differential equation a little later using a simple program.

Taylor and Maclaurin Series

The two techniques we will use to solve differential equations are both based on the Taylor and Maclaurin Series of pure mathematics. These two series relate any function to its derivatives (f'(x), f"(x), f'"(x), etc.). The Taylor series is

$$f(x) = f(a) + ((x-a)/1!)f'(a) + ((x-a)^2)/2!)f"(a) + \ldots + ((x-a)^{n+1}/(n+1)!)f^n(a)+E(x)$$

Without going into details, we just need to see that the series relates the original function f(x) to its derivatives f'(x), f"(x), and so on (where f'(x) = dy/dx and f"(x) = d²y/dx²).

The Maclaurin series is a special case of the Taylor series where a=0 and E(x) = 0. So the Maclaurin series is

$$f(x) = f(0) + xf'(0) + (x^2/2!)f"(0)+\ldots\ldots.$$

Again this just gives the original function in terms of its derivatives and powers of x.

The two methods we will use to solve differential equations use similar relations. One method is called the Euler method and the other is the Runge-Kutta method.

Euler Method

The Euler method, as with the series we have just looked at, relates a function to its derivatives. The relationship between a function and its first derivative is shown in Figure 10-1.

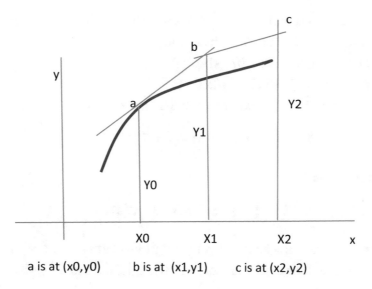

a is at (x0,y0)   b is at (x1,y1)   c is at (x2,y2)

Figure 10-1. *Euler method*

Here, the curved line is our function and the slanted line (ab) which just touches it is the curve gradient at the point where it touches the curve. This gradient is the first derivative evaluated at that point.

Figure 10-2 shows the lines ab and bc. We have projected horizontal lines from a and b and vertical lines from b and c to produce two triangles. For the left-hand triangle, we can see from the graph on the last page that the length of the base must be x1–x0 and the length of the perpendicular line from the base is y1–y0. Doing a similar thing on the right-hand triangle gives us a base of x2–x1 and a perpendicular of y2–y1.

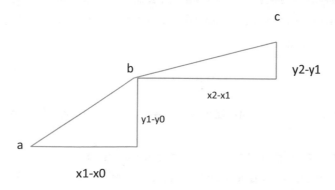

Figure 10-2. *Euler method analysis*

As the gradient of each of these triangles is just the length of the perpendicular side divided by the length of the base, we can say that

Gradient of left triangle = (y1–y0)/(x1–x0)

Gradient of right triangle = (y2–y1)/(x2–x1)

We can relate this to the first two terms of the Taylor series

$$f(x) = f(a) + (x-a)f'(a)$$

which we can rewrite as

$$(f(x) - f(a))/ (x-a) = f'(a)$$

So the left-hand side is the same as our triangle gradients.

If we have the same length of base for each triangle, we can call it h. So we now have

$$(f(x)- f(a))/h = f'(a)$$

or

$$f(x)-f(a) = hf'(a)$$

or

$$f(x) = f(a)+ hf'(a)$$

What we can do with this formula is set initial values of f(x) and a and set up a forloop in which we increase the a value by h for each pass of the loop.

So as we increment our x value, we calculate a new f(x) or y value. So we could write the preceding equation as

$$y_{n+1}= y_n + hf'(x)$$

This is the Euler method of finding f(x), if we already know f'(x) and initial values of f(x) and a.

Following is the code for using the Euler formula in a loop. We have the differentiated function f'(x) and our initial values, and we calculate the value of f(x) after a set number of loops.

```
/*eulermeCh */
#define _CRT_SECURE_NO_WARNINGS
#include<stdio.h>
#include<math.h>
float func(float x, float y);
int main()
{
      float F1;
      float x0, y0, x, y, h, xn;

      printf("Enter initial x, initial y, final x, increment ");
      scanf("%f %f %f %f", &x0, &y0, &xn, &h);

      x = x0;              /* set initial value for loop */
      y = y0;              /* set initial value for loop */
/* forloop contains Euler function */
      for (x = x0;x < xn;x = x)
      {
            F1 = h * func(x, y); /*  h f'(x) from our Euler Formula */

            y = y + F1;   /* y(n+1) = yn + h f'(x) from our Euler Formula */
            x = x + h;    /* increment the x value for the next pass of loop */
            /*printf("X = %f Y = %f\n",x,y);*/
      }
      printf("X = %f Y = %f\n", x, y);

}
float func(float x, float y)
{
      float funcval;
      funcval = 2 * x;     /* Function is dy/dx = 2x */
      return funcval;
}
```

In this program we know that dy/dx = 2x (as you can see in the function func which sets its reply in funcval to the answer 2*x). So the differential equation we are trying to solve here is dy/dx = 2x. This is an easy equation to solve by ordinary calculus, and we

239

should get the answer y = x². If we run this program, we are prompted for initial values of x and y. Set these both at zero. Then we are asked for the final value of x from our loop. Set this to 2. Finally we are asked for the increment of x for each pass of the loop. Set this to 0.1. If we do all this, then the program should print out the final values of x and y. It should give x=2.0 and y=3.800001.

If you consider our calculus answer of y=x², then for x=2 we will get y=4. So our answer of 3.800001 is not too far away. Maybe if we try to have a smaller x increment, we might get closer to 4. Try an increment of 0.01 instead of 0.1. You should get y=4.019997 which is a lot closer to 4.

What we can do is write our data points to a file and then we can write the points to a graph and compare it to the function to see how close we are.

The following code does this and writes to the file "euler1.dat".

```
/*eulermeCh2 */
#define _CRT_SECURE_NO_WARNINGS
#include<stdio.h>
#include<math.h>
float func(float x, float y);
int main()
{
    float F1;
    float x0, y0, x, y, h, xn;

    FILE *fptr;
    fptr = fopen("euler1.dat", "w");

    printf("Enter initial x, initial y, final x, increment ");
    scanf("%f %f %f %f", &x0, &y0, &xn, &h);
    x = x0;
    y = y0;

/* forloop contains Euler function */

    for (x = x0;x < xn;x = x)
    {
        F1 = h * func(x, y);

        y = y + F1;
        x = x + h;
```

```
        fprintf(fptr, "%f\t%f\n", x, y);
        /*printf("X = %f Y = %f\n",x,y);*/
    }
    printf("X = %f Y = %f\n", x, y);

    fclose(fptr);
}
float func(float x, float y)
{
    float funcval;
    funcval = 2 * x; /* Function is dy/dx = 2x */

    return funcval;
}
```

If you import this file as a "point series" file to the Graph package then display it, you can then draw the graph y=x² on the same graph and compare them.

Figure 10-3 is what you should get. The red dots are the points from our program and the continuous blue curve is the function y=x².

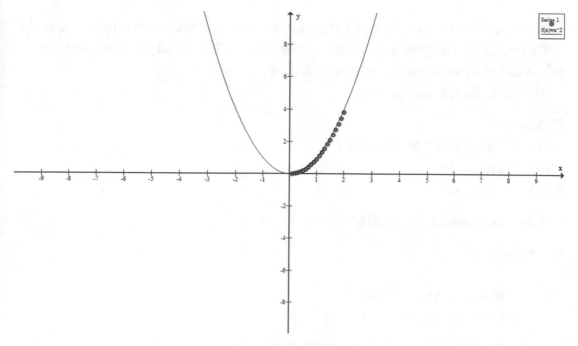

Figure 10-3. *Euler program compared with y=x²*

You can write other functions to use with the Euler method. As you can see from the code, it is only the function, func, in the program which contains this so you can change this from 2*x to other functions.

Runge-Kutta Method

This is a more accurate method than the Euler method. Its formula is a little more complicated than the one for the Euler method, but it's fairly easy to program.

It is really just a more accurate extension to Euler. The formula is

$$y_{x+1} = y_x + (1/6)*(F1 + 2F2 + 2F3 + F4)$$

Where

$$F1 = hf(x,y)$$
$$F2 = hf(x+(h/2),y+(F1/2))$$
$$F3 = hf(x+(h/2),y+(F2/2))$$
$$F4 = hf(x+h,y+F3)$$

So basically we are just using our hf(x,y) term from our Euler method and extending it. We just set up a loop for this function, as we did with Euler, and just increment our values of x and y as we proceed through the loop.

The code for this is as follows.

```
/*rkch1 */
#define _CRT_SECURE_NO_WARNINGS
#include<stdio.h>
#include<math.h>

double func(double x, double y);

int main()
{
    double F1, F2, F3, F4;
    double x0, y0, x, y, h, xn;

    printf("Enter initial x, initial y, final x, increment ");
    scanf("%lf %lf %lf %lf", &x0, &y0, &xn, &h);
```

```
    x = x0;
    y = y0;

/* forloop contains Runge-Kutta function */

    for (x = x0;x < xn;x = x)
    {
    /* Set the values of F1,F2,F3 and F4 from the formula */
        F1 = h * func(x, y);
        F2 = h * func(x + h / 2.0, y + F1 / 2.0) ;
        F3 = h * func(x + h / 2.0, y + F2 / 2.0);
        F4 = h * func(x + h, y + F3);

        /* Increment our y value using the formula*/
        y = y + (F1 + 2 * F2 + 2 * F3 + F4) / 6.0;

        /* Increment the x-value by our chosen entered value */
        x = x + h;

        /* Use the following lone of code (currently commented out) */
        /* If you want to monitor the progress of the forloop */
        /*printf("X = %lf Y = %lf\n",x,y);*/

    }
    printf("X = %lf Y = %lf\n", x, y);

}
double func(double x, double y)
{
    double funcval;
    funcval = 3*pow(x,2); /* function is dy/dx = 3x² */
    return funcval;
}
```

This program uses the function $dy/dx=3*x^2$. Again, we only have to change that function within func if we want to solve a different problem.

If you run this program with initial values x0=0, y0=0, xn=5, and increment=0.1, you should get the result x=5.099998 and y=132.650909.

Again, in this case, we know what the solution to the differential is by using calculus. It is $y=x^3$ so when $x=5$, y should be 125. Again, we could try to make the increment smaller. If we try an increment of 0.01, we get 125.000137 which is very close to the correct answer.

We can compare our points with the correct curve by writing our points to a file. The code for this is as follows.

```c
/*rkch2 */
#define _CRT_SECURE_NO_WARNINGS
#include<stdio.h>
#include<math.h>
double func(double x, double y);
int main()
{
    double F1, F2, F3, F4;
    double x0, y0, x, y, h, xn;

    FILE *fptr;
    fptr = fopen("runge1.dat", "w");/* ! FILE NAME ! */

    printf("Enter initial x, initial y, final x, increment ");
    scanf("%lf %lf %lf %lf", &x0, &y0, &xn, &h);
    x = x0;
    y = y0;

/* forloop contains Runge-Kutta function */

    for (x = x0;x < xn;x = x)
    {
        F1 = h * func(x, y);
        F2 = h * func(x + h / 2.0, y + F1 / 2.0);
        F3 = h * func(x + h / 2.0, y + F2 / 2.0);
        F4 = h * func(x + h, y + F3);
        y = y + (F1 + 2 * F2 + 2 * F3 + F4) / 6.0;
        x = x + h;
        fprintf(fptr, "%lf\t%lf\n", x, y);
```

```
        /*printf("X = %lf Y = %lf\n",x,y);*/
    }
    printf("X = %lf Y = %lf\n", x, y);

    fclose(fptr);
}
double func(double x, double y)
{
    double funcval;
    funcval = 3*pow(x,2); /* function is dy/dx =3 x² */
    return funcval;
}
```

If we write the points to a file and compare it to the curve y=x³, we can see how close our program is (Figure 10-4).

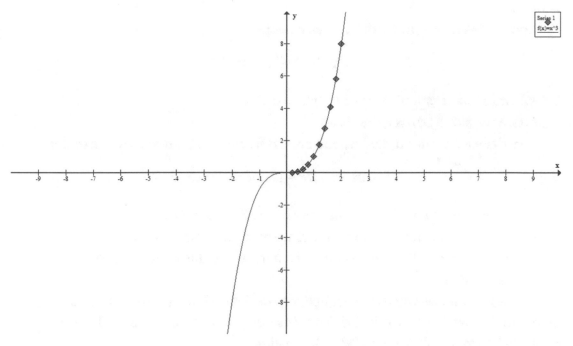

Figure 10-4. *Runge-Kutta program compared with y=x³*

Second Order Differential Equations

The differential equations we have been looking at up to now are called "First Order" differential equations. This means that the original function has only been differentiated once. So, for the function $y = 2x^3$, this differentiates to

$$dy/dx = 6x^2$$

So this is a first order differential equation. If we differentiate this again, we get

$$d^2y/dx^2 = 12x$$

This is a second order differential equation because it contains a term which has been differentiated twice.

To solve a second order differential equation, we do a trick whereby we rewrite the second order differential equation as a combination of two first order differential equations.

Look at this second order differential equation.

$$d^2y/dx^2 = dy/dx - 6x^2 + 12$$

We introduce a variable $v(x)$ where $v(x) = dy/dx$.
So we can say $dv(x)/dx = d^2y/dx^2$.
So we can write our original second order differential equation as two equations.

$$dv/dx = v - 6x^2 + 12 \text{ and } dy/dx = v$$

Now, if we are given initial conditions when x=0, y=0, and $dy/dx = v = 0$, we can use our Runge-Kutta method, we do a trick we used earlier in this chapter for finding solutions to first order differential equations, and just modify them to solve our two equations together.

In our program we will use two functions, one for each of our two first order differential equations. One is called func1 (for our $dy/dx = v$ equation) and the other is called func2 (for our $dv/dx = v - 6x^2 + 12$ equation).

In our loop to perform the four Runge-Kutta stages using F1, F2, F3, and F4, we will keep the two calculations for the two equations separate and rename these four terms D1F1, D1F2, D1F3, and D1F4 for the first set and D2F1, D2F2, D2F3, and D2F4 for the second set. Our initial values are entered by the user (although you can preset these in your

246

program if you wish). The initial values are the initial value of x, the initial value of y, the initial value of dy/dx (labeled here as y2), the final value of x, and the increment value.

We will use a function that we already know its form so that we can compare our answer with the correct one. We can, again, use the Graph package to draw the correct graph and then import the point series, from the output of our program, to the same graph.

We will solve the second order differential equation $d^2y/dx^2 = dy/dx - 6x^2 + 12x$. We are using an equation to which we already know the solution. It is $y = 2x^3$.

So we can say that $dy/dx = 6x^2$ or $y2 = 6x^2$.

So our initial conditions are when x = 0, y = 0, and when x = 0 y2 (=dy/dx) = 0.

We can try an increment of 0.1 as a starter then reduce it to, say, 0.01 for our second attempt.

The following is the code for our second order differential equation solution.

```
/*runge2me5a */
/* Second Order Differential Equation */
/* Solves d2y/dx2  = dy/dx - 6x^2 + 12x */
/* Splits the second order DE into two first order DEs */
#define _CRT_SECURE_NO_WARNINGS
#include<stdio.h>
#include<math.h>

double y1a,y2;

double func1(double x,double y);
double func2(double x,double y);

int main()

{
      double D1F1,D1F2,D1F3,D1F4;
      double D2F1,D2F2,D2F3,D2F4;
      double x0,x,h,xn;

      FILE *fptr1;

      printf("Enter initial x, initial y1,initial y2 ,final x, increment ");
      scanf("%lf %lf %lf %lf %lf",&x0,&y1a,&y2,&xn,&h) ;
      x=x0;
```

```
        fptr1=fopen("rk25.dat","w");

        for(x=x0;x<xn;x=x)
        {
                /* Derivative for 1st first order DE */

                D1F1=h*func1(x,y1a);
                D1F2=h*func1(x+h/2.0,y1a+D1F1/2.0);
                D1F3=h*func1(x+h/2.0,y1a+D1F2/2.0);
                D1F4=h*func1(x+h,y1a+D1F3);
                y1a=y1a+(D1F1+2*D1F2+2*D1F3+D1F4)/6.0;

                /* Derivative for 2nd first order DE */

                D2F1=h*func2(x,y2);
                D2F2=h*func2(x+h/2.0,y2+D2F1/2.0);
                D2F3=h*func2(x+h/2.0,y2+D2F2/2.0);
                D2F4=h*func2(x+h,y2+D2F3);
                y2=y2+(D2F1+2*D2F2+2*D2F3+D2F4)/6.0;

                x=x+h;
                /*printf("X = %f Y = %f\n",x,y);*/
                fprintf(fptr1,"%f\t%f\n",x,y1a);
        }

        printf("X = %lf Y = %lf\n",x,y1a);
        fclose(fptr1);

}
double func1(double x,double y)
{
        /* First order differential equation no. 1 */
        double funcval;
        funcval=y2;

        return funcval;
}
```

```
double func2(double x,double y)
{
    /* First order differential equation no. 2 */
    double funcval;

    funcval=y2-6*pow(x,2)+12*x;

    return funcval;
}
```

The forloop contains the two sets of Runge-Kutta terms, one for each of the two equations. The first set deals with the first of our first order differential equations and calls the function func1, and the second set deals with the second of the first order differential equations and calls the function func2.

Output data is written to the file rk25.dat. The comparison of this data with the expected curve of $y = 2x^3$ is shown in Figure 10-5. The red curve is the Runge-Kutta data produced by the program.

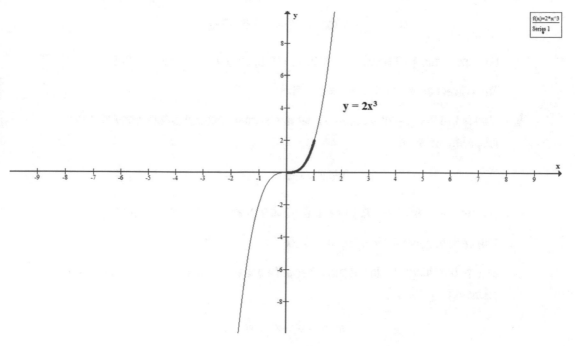

Figure 10-5. *Runge-Kutta program compared with $y=2x^3$*

EXERCISES

1. Amend your Euler program which writes to an output file to solve the
   differential equation $dy/dx = (-1)/x^2$. Use initial values $x0 = 1$, $y0 = 1$, $xn = 5$,
   $inc = 0.1$.

   Repeat Q1 for the differential equation $dy/dx = 4x/(5-x^2)^2$. Use initial values
   $x0 = 0$, $y0 = 0.4$, $xn = 2$, $inc = 0.1$.

2. Amend your Runge-Kutta program to solve the differential equation
   $dy/dx = 1/sqrt(x^2 + y^2)$. Use initial values $x0 = 0$, $y0 = 1$, $xn = 1$, $inc = 0.25$.

3. Solve the differential equation shown at the start of this chapter. It was
   $dy/dx = (4x - 6x^2) / exp(3x)$. Use initial values $x0 = 0$, $y0 = 0$, $xn = 1$,
   $inc = 0.005$.

4. Modify the Runge-Kutta second order differential equation program to solve the
   following equation.

   $$d^2y/dx^2 = dy/dx - 12x + 12$$

   For input values $x0 = 0.0$, $y0 = 1.0$, $y2 = 0.0$, $xn = 1.0$, and $inc = 0.005$.

   The correct curve for this is $y = e^x + 6x^2$.

5. Modify the Runge-Kutta second order differential equation program to solve the
   following equation.

   $$d^2y/dx^2 = dy/dx - 3x^2 + 6$$

   For input values $x0 = 0.0$, $y1a = 0.0$, $y2 = 0.0$, $xn = 2.0$, and $h = 0.005$.

   The correct curve for this is $y = x^3 + 3x^2$.

6. Modify the Runge-Kutta second order differential equation program to solve the
   following equation.

   $$d^2y/dx^2 = dy/dx + 5y + e^{-2x}$$

   For input values $x0 = 0.0$, $y1a = 1.0$, $y2 = -2.0$, $xn = 1.0$, and $h = 0.005$.

   The correct curve for this is $y = e^{-2x}$.

APPENDIX A

Development Environment Reference

Visual Studio

The following is a screenshot of Visual Studio. If you are not familiar with this, the upper-middle box shows the source code and the lower-middle box shows runtime messages.

© Philip Joyce 2019

P. Joyce, *Numerical C*, https://doi.org/10.1007/978-1-4842-5064-8

Debugging is relatively easy in Visual Studio. On the upper-middle box which is displaying your code, you can select breakpoints. These are positions in the code where you want to pause your program when it runs so that you can examine integers, arrays, and so on in your program to see if they contain the values you are expecting. What you need to do to set breakpoints is to click the line where you want to stop on (click the blue bar to the left of the box on the line). You can see the three breakpoints for this program. They are at the positions of the three red dots.

Then press F5 to start your program.

In the lower-middle box, the first breakpoint that your program reaches will be displayed.

Move your cursor over the line of code, and it will display the contents of the variable you are moving the cursor over (e.g., $r1 = 4$ or $c1 = 5$).

Press F5 again to continue your program (or "continue" on "debug" on the toolbar).

If you hover your cursor over the red dot or the arrow inside the dot on the blue bar to the left of the code, it shows you the line number.

You can also enter the name of the variables you wish to look at on the lower-middle box so that you can see them changing as you move through your breakpoints. You can also click "locals" in this box, and it will display them automatically.

If you want to find a line number in your source code, you click "Edit" on the toolbar, then hover over "Goto". Move to "Go to Line" then type in the line number.

When you run your program, if the output window closes before you have seen the output, you can do one of two things:

- Right-click the Project name, then expand Configuration Properties ➤ Linker ➤ system. Then subsystem ➤ console.

- You can insert the line of code system("pause"); as the last executable line in your program.

If you use one of these, then the window will stay open until you press a key. Let's try the first one:

1.   Right-click the Project name.

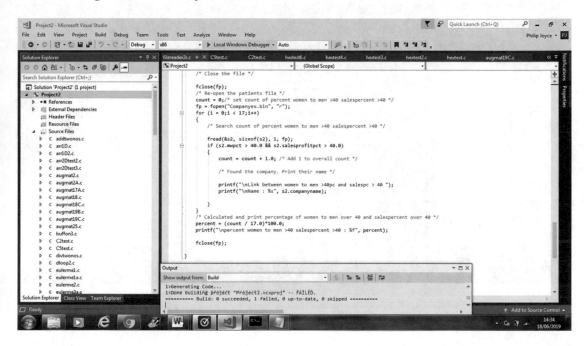

2.   Click Properties.

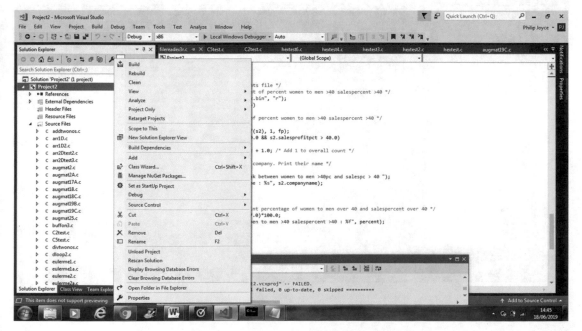

3. Click Configuration Properties.

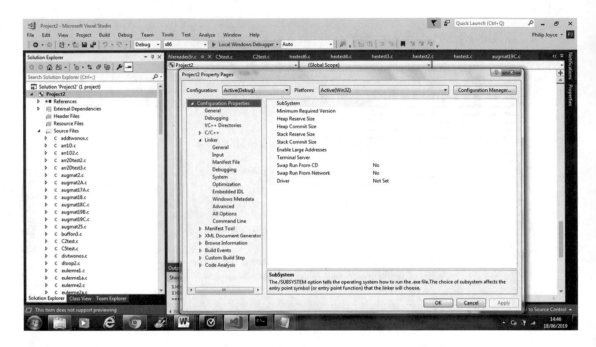

4. Click Linker.

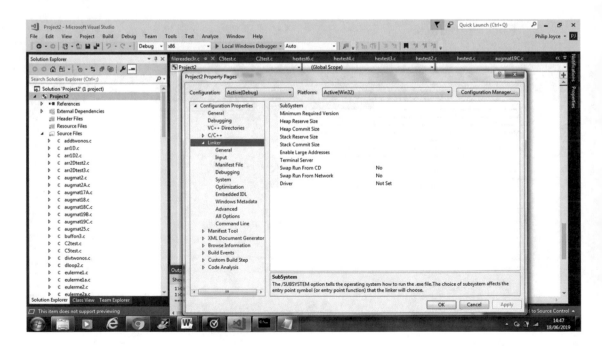

5. Click System then SubSystem.

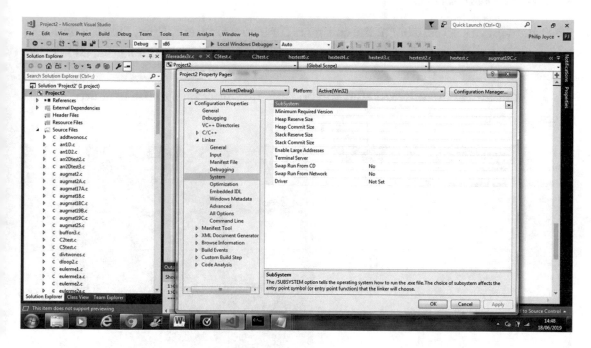

6. Click the down arrow at the right of SubSystem.

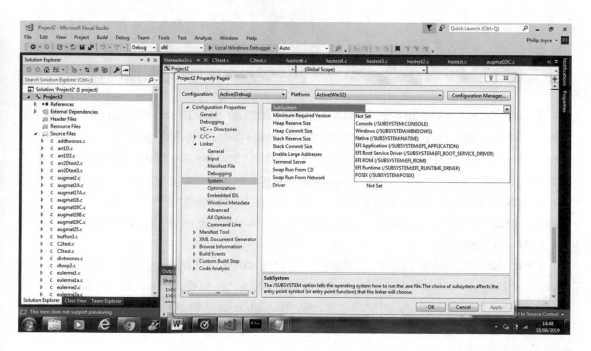

7. Click Console then OK.

Command Line

The following is a screenshot of the command line environment. Here, only the dark box to the left is the command line environment. You can display your source code on a separate window.

All commands are entered after the prompt and messages to the user are displayed there.

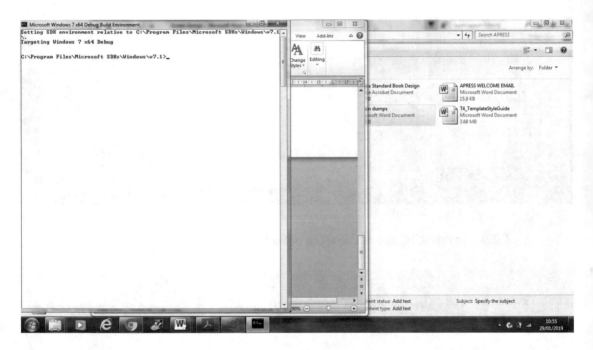

If you look at the command line in the following figure, you can see the prompt being shown. The computer is waiting for the user to enter data. The "c:\" just shows you the directory you are in. If you type "dir" you get a list of what is currently contained in the directory. If you look halfway up the screen, you will see that this has been entered and the list is shown. Subdirectories (directories within the current directory) are shown by "<DIR>" next to them. Anything which is not a directory will be a file, and the size of the file is shown next to it.

If you want to create a new directory from your current directory, you type in "md name" where "name" is the name you want to call your new directory. You can then move to your new directory by typing "cd name". Once in your new directory, you can move back to your previous one by typing "cd ..". The prompt always shows the current directory you are in.

You compile and test from the command line.

To compile, type in "cc progname.c".

To run the program, just type the program name without the ".c".

Any output from your program will appear on the command line.

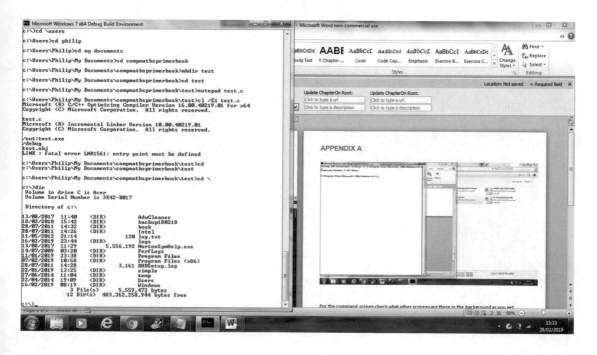

APPENDIX B

Syntax Reference

Mathematical Functions

```
double sin(double x) input x in radians - returns sine
double cos(double x) input x in radians - returns cosine
double tan(double x) input x in radians - returns tangent
double asin(double x) input x -  returns arcsin in radians
double acos(double x) input x -  returns arccos in radians
double atan(double x) input x  - returns arctan in radians
double sinh(double x) input x - returns hyperbolic sine
double cosh(double x) input x - returns hyperbolic cosine
double tanh(double x) input x - returns hyperbolic tangent
double exp(double x) input x - returns e to power x
double log(double x) input x - returns natural log of x
double log(double x) input x - returns log to base 10 of x
double pow(double x, double y)- returns x to power y
double sqrt(double x)- returns square root of x
double ceil(double x)- returns smallest integer >= x
double fabs(double x)- returns absolute value of x
double floor(double x)- returns largest integer <= x
double fmod(double x, double y)- returns remainder of x / y
double modf(double x, double *int)- returns part of x after    decimal point
sets int to integer part
```

Pointers

We have seen the use of pointers in C when writing out file access programs. Our variables that we declare at the start of our programs reserve an area of the computer store which we can use during our program's operation. Each time we run our program,

© Philip Joyce 2019
P. Joyce, *Numerical C*, https://doi.org/10.1007/978-1-4842-5064-8

it will be loaded into a different area of store. As computers have enormous areas of store, we use hexadecimal numbers as addresses of the parts of store we are using. (Hexadecimal numbers are numbers to base 16 rather than to base 10 that we use with decimal numbers – so 10 in decimal would be ten, but 10 in hexadecimal would be sixteen. A typical hexadecimal number could be 4ef20a5.)

Here is a program to illustrate the basic meaning of pointers.

```
/* AppAptr1 */
#include <stdio.h>

int main()
{
      int ourvariable;
      char achar;
      char anarray[10];

printf("address of ourvariable is %p\n",&ourvariable);
printf("address of achar is %p\n",&achar);
printf("address of anarray is %p\n",&anarray);
return(0);
}
```

Notice that in printf we use %p to identify the output as a pointer at an address which is a hexadecimal number.

When you run this program, you will get output something like

>   address of ourvariable is 000000000019f728

>   address of achar is 000000000019f72c

>   address of anarray is 000000000019f719

If you run the program again, you may get something like

>   address of ourvariable is 000000000029fe98

>   address of achar is 000000000029fe9c

>   address of anarray is 000000000029fe88

So the addresses are different each time you run the program as your program will have been loaded into a different part of the computer's store.

What we want to show here is what the pointers are.

Notice that in our printf commands, we used &ourvariable etc as our parameter. This means that you want to print the address of that variable, not its contents. So in each of the printf commands, we do the same thing for each variable and so we get the address of each variable.

Look at the next program to see how we can make use of this.

```c
/* AppAptr2 */
#include <stdio.h>

int main()
{
        int ourvariable = 38;
        char achar = 'M';
        char anarray[10] = "HELLO";

        int *ourvariablep;
        char *acharp;
        char *anarrayp;

        ourvariablep = &ourvariable;
        acharp = &achar;
        anarrayp = anarray;

        printf("address of ourvariable is %p\n",&ourvariable);
        printf("value in ourvariable is %d\n",ourvariable);
        printf("address in ourvariablep is %p\n",ourvariablep);

        printf("address of achar is %p\n",&achar);
        printf("value in achar is %c\n",achar);
        printf("address in acharp is %p\n",acharp);

        printf("address of anarray is %p\n",&anarray);
        printf("value in anarray is %s\n",anarray);
        printf("address in anarrayp is %p\n", anarrayp);

        return(0);
}
```

Here we preset an int type of variable, a char type, and a char array. We declare pointer variables for each of our three preset variables, then we print out what is in each. When we run this, we would get something like this:

> address of ourvariable is 000000000021fb08
> value in ourvariable is 38
> address in ourvariablep is 000000000021fb08
>
> address of achar is 000000000021fb18
> value in achar is M
> address in acharp is 000000000021fb18
>
> address of anarray is 000000000021faf0
> value in anarray is HELLO
> address in anarrayp is 000000000021faf0

Standard Library Functions

stdio.h   (Input/Output)

getchar()   – returns character typed in

putchar()   – prints character to screen

scanf()   – reads a set of characters typed in

printf()   – prints a set of characters to screen

fgets()   – returns a string typed in

fputs()   – writes a string to the screen

math.h (Mathematical Functions)

as described earlier

string.h   (String Functions)

strlen() – returns length of string

strcmp() – compares two strings

strcpy() – copies second string to first

strcat() – concatenates second string to first

stdlib.h()

srand() – initializes starting point for rand() calls

rand() – returns a random number between 0 and 1

malloc() – dynamically allocates store area to program

free() – frees the storage allocated by malloc()

Comparing Double, Float, and Integer

```
/* Program to show differences in accuracy of arithmetic values between
double float and integer */
/* AppAcomp */
#include<stdio.h>
main()
{
    float f,f1,f2;
    double d,d1,d2;
    int i,i1,i2;
    /* We want to divide 1623875 by 57 in double format, float format and
    integer format */

    f1=1623875;
    f2=57;

    d1=1623875;
    d2=57;
```

```
    i1=1623875;
    i2=57;

    d=d1/d2;
    f=f1/f2;
    i=i1/i2;

    printf("d is %lf, f is %f, i is %d\n",d,f,i);

/* Answer to this is d = 28489.035088 f = 28489.035156 i = 28489
    Calculator anwer is 28489.035087719289245614 (recurring)
*/

}
```

APPENDIX C

Answers to Problems

CHAPTER 1

1. (i) 289 (ii) 10303 (iii) −10303 (iv) 1000

2. (i) 1.3 (ii) 1

3. (i) 194.56 (ii) −12 (iii) −166064.5125 (iv) −5385.81108 (v) 0

4. (i) 3 (ii) 205 (iii) 1.528118393

5. 1.528118393

6. The code is as follows.

```
#define _CRT_SECURE_NO_WARNINGS
#include<stdio.h>
/* demonstrate a forloop (setting the forloop limit)*/
main()

{

    float this_is_a_number ,  total;
    int i,forlimit;

    total = 0;
    printf( "Please enter forloop limit:\n " );
        scanf( "%d", &forlimit );/* entered limit stored in
        forlimit */
    for(i=0;i<forlimit;i++)
```

© Philip Joyce 2019
P. Joyce, *Numerical C*, https://doi.org/10.1007/978-1-4842-5064-8

```
        {
                printf( "Please enter a number:\n " );
                scanf( "%f", &this_is_a_number );
                total = total + this_is_a_number;

        }
        printf("Total Sum is = %f\n",total);

    }
```

7. 60

8. Should only perform the do loop on the first pass. On the first test (after the first pass), it is greater. If you set it to 10, it will go on forever (press CTRL+C to abort it).

9. The program for this is as follows.

```
#define _CRT_SECURE_NO_WARNINGS

#include<stdio.h>

/* example of a 2D array test for 2 arrays*/
int main()
{
        int arr1[8][8];
        int arr2[8][8];

        int i,j,k,l;

        printf("enter number of rows and columns of first array(max 8
        rows max 8 columns) \n");
        scanf("%d %d", &k, &l);
        if(k>8 || l>8)
        {
                printf("error - max of 8 for rows or columns\n");

        }

        else
        {
                printf("enter array\n");
                for(i=0;i<k;i++)
```

```c
    {
        for(j=0;j<l;j++)
        {
                scanf("%d",&arr1[i][j]);
        }
    }
    printf("Your array is \n");
    for(i=0;i<k;i++)
    {
        for(j=0;j<l;j++)
        {
                printf("%d ",arr1[i][j]);
        }
    printf("\n");

    }
}
printf("first row of first array\n");
for(j=0;j<k;j++)
{
     printf("%d ",arr1[0][j]);
}

printf("enter number of rows and columns of second array(max 8
rows max 8 columns) \n");
scanf("%d %d", &k, &l);
if(k>8 || l>8)
{
     printf("error - max of 8 for rows or columns\n");

}

else
{
     printf("enter array\n");
     for(i=0;i<k;i++)
     {
        for(j=0;j<l;j++)
        {
```

```
                                scanf("%d",&arr2[i][j]);
                    }
            }
            printf("Your array is \n");
            for(i=0;i<k;i++)
            {
                    for(j=0;j<l;j++)
                    {
                            printf("%d ",arr2[i][j]);
                    }
                    printf("\n");

            }
        }
        printf("first row of second array\n");
        for(j=0;j<k;j++)
                    {
                            printf("%d ",arr2[0][j]) ;
                    }

            printf("\n");
    }
```

10. A program to do this is as follows.

```
/* Function which returns an answer  */
/* finds the pupil in one year of the school with the highest marks */

#define _CRT_SECURE_NO_WARNINGS
#include <stdio.h>
double getmarks(double pupils[]);

int main()
{
    double pupil;
    /* Array with marks for class is preset in the main part of the
    program */
    double  marks[] = {1.2,2.3,3.4,4.5,5.6,6.7,7.8,8.9,9.0};
    /* Call function getmarks. The function returns the average
    marks which is then stored in pupil */
```

```
        pupil = getmarks(marks);
        printf("Avarage mark is   = %lf", pupil);
        return 0;
}

double getmarks(double pupils[])
{
        int i;
        double average, total;
        total = 0;
        /* Go through all the pupils in turn and add their mark */
        for (i = 0; i < 9; ++i)
        {
                total = total + pupils[i];

        }
        average = total/9;
        return average; /* returns the value in average to where the
        function was called */
}
```

11. 6.062177826 m

12. 12.74558747

13. 22.03597347

14. The code for this is as follows.

```
/* Structure example program (extended structure)*/
#define _CRT_SECURE_NO_WARNINGS
#include<stdio.h>

/* define the structure */
struct Student {
   int id;
   char name[16];
   float percent;
};
```

```c
int main() {
    int i;
/* define 5 data locations of type "student" */

    struct Student year9[5];

    for(i=0; i<5; i++)
    {
        /* Assign values to the structure */
        printf("enter student ID\n");
        scanf("%d",&year9[i].id);
        printf("enter student name\n");
        scanf("%s",year9[i].name);
        printf("enter student percent\n");
        scanf("%f",&year9[i].percent);

    }
    for(i=0; i<5; i++)
    {
        /* Print out structure s1 */

        printf("\nid : %d", year9[i].id);
        printf("\nName : %s", year9[i].name);
        printf("\nPercent : %f", year9[i].percent);

    }

    return (0);
}
```

CHAPTER 2

1. Solution 2   Upper 1.5   Lower 2.5

   Solution 4   Upper 4.5   Lower 3.5

2. Solution 2   Upper 1.5   Lower 5.5

   Solution 6   Upper 6.5   Lower 5.5

3. Solution −2.750075   Upper −2.2   Lower −3.4

    Solution −0.384249   Upper −1   Lower 0

    Solution 2.750075   Upper 2.9   Lower 2.6

4. Solution −1.640314   Upper −2   Lower −1

    Solution 1.580737   Upper 2   Lower 1

5. Solution −1.709976   Upper −1   Lower −2

6. Solution −3.645751   Upper −4   Lower −3

    Solution 1.645751   Upper 2   Lower 1

7. Complex Solution

8. low   0.9   high   1.1   answer 1.0

The following is a screenshot of the function.

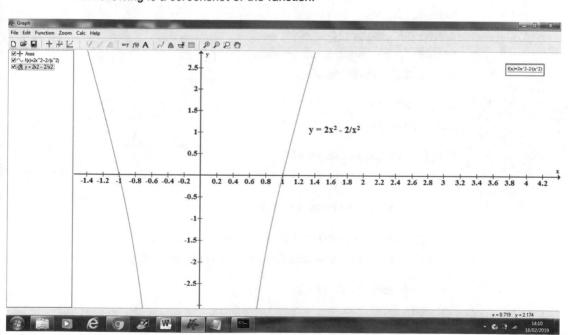

Example code for this question is as follows.

```c
#define _CRT_SECURE_NO_WARNINGS
#include <stdio.h>
#include <math.h>
main()
{

    float lower,upper;
    int i;
    double testhigh,testlow,testvalue,middle;
    int iterations;

    printf("enter lower limit");/* the lower x value for your
    integration */
    scanf("%f",&lower);
    printf("enter upper limit");/* the upper x value for your
    integration */
    scanf("%f",&upper);
    printf("enter number of iterations");
    scanf("%d",&iterations);

    testlow=lower;
    testhigh=upper;

    for(i=0;i<iterations;i++)
    {

        middle=(testhigh+testlow)/2;

        /* sets testvalue to 2*(middle)2 -2/(middle)2 */
        testvalue=2*pow(middle,2)-2*pow(middle,-2);

        if(testvalue == 0)
        {
            printf("x is %f",middle);
            return(0);
        }
        if(testvalue > 0)
        {
            testhigh=middle;
        }
```

```
        else
        {
                testlow=middle;
        }

    }
    printf("x is %f",middle);
}
```

9. low   1.2   high   0.8   answer 0.987175

The following is a screenshot of the function.

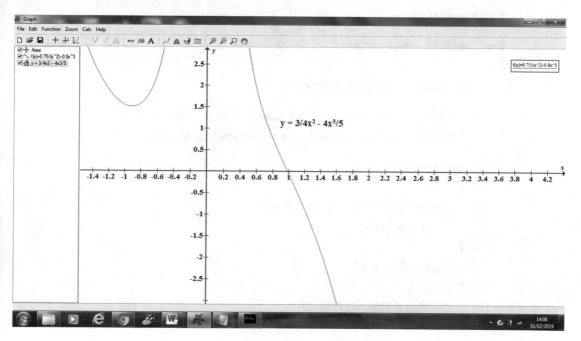

Example code for this question is as follows.

```
/*trapezium -  trial and improvement using inverse functions */
#define _CRT_SECURE_NO_WARNINGS
#include <stdio.h>
#include <math.h>
main()
{
    float lower,upper;

    int i;
```

```c
double testhigh,testlow,testvalue,middle;

int iterations;

printf("enter lower limit");/* the lower x value for your integration */
scanf("%f",&lower);
printf("enter upper limit");/* the upper x value for your integration */
scanf("%f",&upper);
printf("enter number of iterations");
scanf("%d",&iterations);

testlow=lower;
testhigh=upper;

for(i=0;i<iterations;i++)
{
     middle=(testhigh+testlow)/2;

     /* sets testvalue to 0.75/(middle)² -0.8*(middle)³ */
     testvalue=0.75*pow(middle,-2)-0.8*pow(middle,3);

     if(testvalue == 0)
     {
          printf("x is %f",middle);
          return(0);
     }
     if(testvalue > 0)
     {
          testhigh=middle;
     }
     else
     {
          testlow=middle;
     }

}
printf("x is %f",middle);

}
```

10.   low   0.9   high   0.7   answer   0.837863

The following is a screenshot of this function.

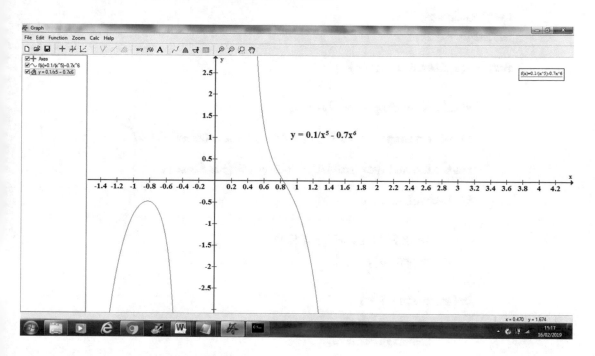

Example code for this question is as follows.

```
/*trapezium -   trial and improvement using inverse functions */
#define _CRT_SECURE_NO_WARNINGS
#include <stdio.h>
#include <math.h>
main()
{
    float lower,upper;

    int i;
    double testhigh,testlow,testvalue,middle;

    int iterations;

    printf("enter lower limit");/* the lower x value for your integration */
    scanf("%f",&lower);
    printf("enter upper limit");/* the upper x value for your integration */
    scanf("%f",&upper);
```

```
printf("enter number of iterations");
scanf("%d",&iterations);

testlow=lower;
testhigh=upper;

for(i=0;i<iterations;i++)
{
    middle=(testhigh+testlow)/2;

    /* sets testvalue to 0.1/(middle)⁵ -0.7*(middle)⁶ */

    testvalue=0.1*pow(middle,-5)-0.7*pow(middle,6);

    if(testvalue == 0)
    {
        printf("x is %f",middle);
        return(0);
    }
    if(testvalue > 0)
    {
        testhigh=middle;
    }
    else
    {
        testlow=middle;
    }

}
printf("x is %f",middle);

}
```

CHAPTER 3

1. 4.410706

2. 4.670777

3. 1.281720

4. 0.583334

5. 1.833334

6. 6.625486

7. 1.164644

The following is a screenshot of this function.

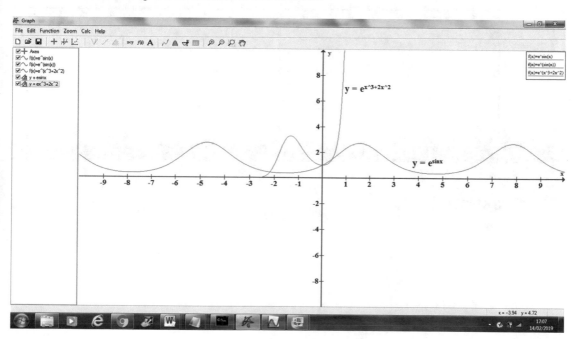

8. 0.632210

9. 0.813667

10. 2.793582

The following is a screenshot of this function.

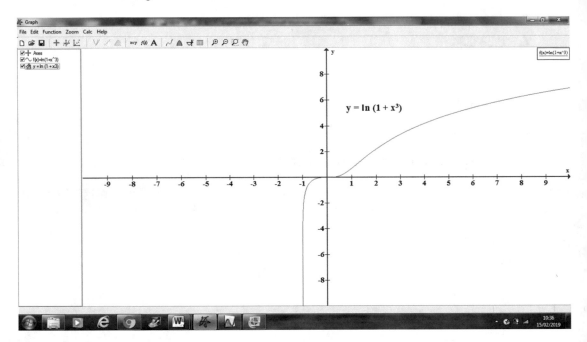

11. 0.874957

12. 0.543001

   The following is a screenshot of this function.

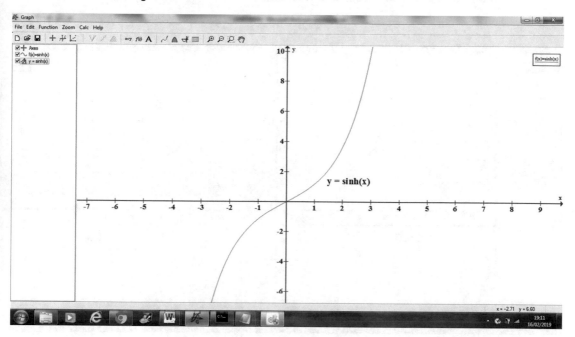

13. 1.175201

   The following is a screenshot of this function.

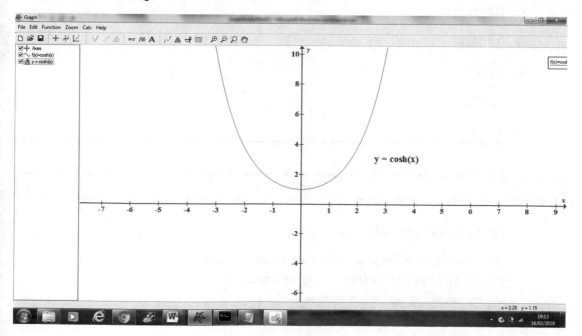

14. 0.433701

The following is a screenshot of this function.

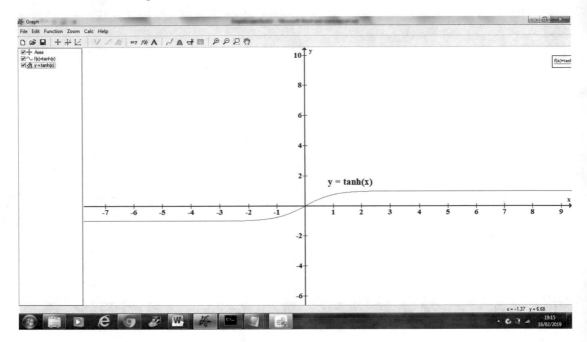

15. 2.544394

16. 0.881943

17. 0.203107

18. 1.597882

19. 1.464231

20. 1.455749

CHAPTER 4

1. Possible code is as follows.

```
/* Montecarlo sphere (whole sphere in 1st quadrant)*/
/* Calculation of volume using monte carlo */
/* by counting relative volumes */
```

```c
/* integrates (x-2)^2 + (y-2)^2 + (z-2)^2 = 2^2 to your specified
limits */
#define _CRT_SECURE_NO_WARNINGS
#include <stdio.h>
#include <stdlib.h>
#include <math.h>
main()
{

    double x, y, z;
    double zupper, zlower, yupper, ylower, xupper, xlower;
    double  montevol, volume;
    double totalexpvol, totalvol;
    int j;
    /*      unsigned int iterations;*/
    long int iterations;

    printf("enter lower x limit\n");
    scanf("%lf", &xlower);
    printf("enter upper x limit\n");
    scanf("%lf", &xupper);
    printf("xlower %lf xupper %lf\n", xlower, xupper) ;

    printf("enter lower y limit\n");
    scanf("%lf", &ylower);
    printf("enter upper y limit\n");
    scanf("%lf", &yupper);
    printf("ylower %lf yupper %lf\n", ylower, yupper);

    printf("enter lower z limit\n");
    scanf("%lf", &zlower);
    printf("enter upper z limit\n");
    scanf("%lf", &zupper);
    printf("zlower %lf zupper %lf\n", zlower, zupper);

    volume = (xupper - xlower)*(yupper - ylower)*(zupper - zlower);
    printf("volume is %lf\n", volume);
    printf("enter iterations up to 1000000\n");
    scanf("%d", &iterations);

    totalvol = 0;
```

281

```
totalexpvol = 0;

for (j = 1;j < iterations;j++)
{

    /* find random numbers for x,y and z */
    x = rand() % 1000;
    y = rand() % 1000;
    z = rand() % 1000;
    y = y / 1000;
    x = x / 1000;
    z = z / 1000;

    /* x,y and z will have numbers between 0 and 1 */
    /* so multiply by the user's entered ranges for x,y and z */
    x = xlower + (xupper - xlower)*x;
    y = ylower + (yupper - ylower)*y;
    z = zlower + (zupper - zlower)*z;

    if (x >= xlower && y >= ylower && z >= zlower)
    {
        totalvol = totalvol + 1; /* This contains the total
        numbe of entries */

        if ((pow((y - 2), 2) + pow((x - 2), 2)) +
        pow((z - 2), 2) < 4)
        {

            totalexpvol = totalexpvol + 1;/* This contains
            number of entries within desired vol */

        }
    }

}
if (totalvol != 0)
{
    montevol = volume * (totalexpvol / totalvol);/* Monte
    Carlo volume os the fraction of the cube volume */
}
printf("monte carlo volume is %lf\n", montevol);

}
```

2. The code is as follows.

```c
/* Montecarlo cone*/
/*      Calculation of volume using monte carlo */
/*      by counting relative volumes */
/* integrates x^2 + y^2 * z to your specified limits */
#define _CRT_SECURE_NO_WARNINGS
#include <stdio.h>
#include <stdlib.h>
#include <math.h>
main()
{

        double x, y, z;
        double zupper, zlower, yupper, ylower, xupper, xlower;
        double  montevol,  volume;
        double totalexpvol, totalvol, tantheta,  radius;
        int j;
        int iterations;

        printf("enter lower x limit\n");
        scanf("%lf", &xlower);
        printf("enter upper x limit\n");
        scanf("%lf", &xupper);
        printf("xlower %lf xupper %lf\n", xlower, xupper);

        printf("enter lower y limit\n");
        scanf("%lf", &ylower);
        printf("enter upper y limit\n");
        scanf("%lf", &yupper);
        printf("ylower %lf yupper %lf\n", ylower, yupper);

        printf("enter lower z limit\n");
        scanf("%lf", &zlower);
        printf("enter upper z limit\n");
        scanf("%lf", &zupper);
        printf("zlower %lf zupper %lf\n", zlower, zupper);

        volume = (xupper - xlower)*(yupper - ylower)*(zupper -
        zlower);/* volume of cuboid enclosing the cone */
```

```c
printf("volume is %lf\n", volume);

printf("enter iterations up to 1000000\n");
scanf("%d", &iterations);

totalvol = 0;
totalexpvol = 0;
tantheta = (zupper - zlower) / (xupper - xlower);/* Tangent of
the angle the slant edge makes with the base */
radius = sqrt(pow(xupper, 2) + pow(yupper, 2));
radius = 2;

for (j = 1;j < iterations;j++)
{
    /* find random numbers for x,y and z */
    x = rand() % 1000;
    y = rand() % 1000;
    z = rand() % 1000;
    y = y / 1000;
    x = x / 1000;
    z = z / 1000;

    /* x,y and z will have numbers between 0 and 1 */
    /* so multiply by the user's entered ranges for x,y and z */
    x = xlower + (xupper - xlower)*x;
    y = ylower + (yupper - ylower)*y;
    z = zlower + (zupper - zlower)*z;

    if (x >= xlower && z >= zlower && y >= ylower)
    {
        totalvol = totalvol + 1; /* This contains the total
        number of entries */
        /* x and y coordinates have to be within circular
        base */
        /* z coordinate has to be below the slanted edge
        which */
        /* is vertically above the (x,y) point */
```

```
                    if ((pow(y, 2) + pow(x, 2) < 4) && (z <
                    tantheta*(radius - sqrt(pow(x, 2) + pow(y, 2)))))

                    {

                            totalexpvol = totalexpvol + 1;/* This contains
                            number of entries within desired vol */

                    }

              }

        }
        if (totalvol != 0)
        {
              montevol = volume * (totalexpvol / totalvol);/* Monte
              Carlo volume os the fraction of the cube volume */
        }
        printf("monte carlo volume is %lf\n", montevol);

}
```

3. The code is as follows.

```
/* Montecarlo 4-D sphere*/
/* Calculation of volume using monte carlo */
/* by counting relative volumes */
/* integrates x^2 + y^2 + z^2 + p^2= 2^2 to your specified limits */
/* NB 4D graphs have 16 "quadrants"(8 for 3D, 4 for 2D) */
#define _CRT_SECURE_NO_WARNINGS
#include <stdio.h>
#include <stdlib.h>
#include <math.h>
main()
{

        double x, y, z, p;
        double zupper, zlower, yupper, ylower, xupper, xlower, pupper,
        plower;
        double  montevol, volume;
        double totalexpvol, totalvol;
        int j;
        int iterations;
```

```c
printf("enter lower x limit\n");
scanf("%lf", &xlower);

printf("enter upper x limit\n");
scanf("%lf", &xupper);
printf("xlower %lf xupper %lf\n", xlower, xupper);

printf("enter lower y limit\n");
scanf("%lf", &ylower);
printf("enter upper y limit\n");
scanf("%lf", &yupper);
printf("ylower %lf yupper %lf\n", ylower, yupper);

printf("enter lower z limit\n");
scanf("%lf", &zlower);
printf("enter upper z limit\n");
scanf("%lf", &zupper);
printf("zlower %lf zupper %lf\n", zlower, zupper);

printf("enter lower p limit\n");
scanf("%lf", &plower);
printf("enter upper p limit\n");

scanf("%lf", &pupper);
printf("plower %lf pupper %lf\n", plower, pupper);
volume = (xupper - xlower)*(yupper - ylower)*(zupper -
zlower)*(pupper - plower);
printf("volume is %lf\n", volume);

printf("enter iterations up to 1000000\n");
scanf("%d", &iterations);

totalvol = 0;
totalexpvol = 0;

for (j = 1;j < iterations;j++)
{
        /* find random numbers for x,y ansd z */
        x = rand() % 1000;
        y = rand() % 1000;
```

```
        z = rand() % 1000;
        p = rand() % 1000;
        y = y / 1000;
        x = x / 1000;
        z = z / 1000;
        p = p / 1000;
        /* x,y and z will have numbers between 0 and 1 */
        /* so multiply by the user's entered ranges for x,y and z */
        x = xlower + (xupper - xlower)*x;
        y = ylower + (yupper - ylower)*y;
        z = zlower + (zupper - zlower)*z;
        p = plower + (pupper - plower)*p;

        if (x >= xlower && z >= zlower && y >= ylower && p >= plower)
        {
                totalvol = totalvol + 1; /* This contains the total
                number of entries */

                if ((pow(y, 2) + pow(x, 2) + pow(z, 2) + pow(p, 2)) < 4)
                {
                        totalexpvol = totalexpvol + 1;/* This contains
                        number of entries within desired vol */

                }
        }

}
if (totalvol != 0)
{
        montevol = volume * (totalexpvol / totalvol);/* Monte
        Carlo volume os the fraction of the cube volume */
}
printf("monte carlo volume is %lf\n", montevol);

}
```

CHAPTER 5

1. Program should be as follows.

```
/* Matrix program */
/* Add two floating point matrices */
    #define _CRT_SECURE_NO_WARNINGS

#include<stdio.h>

main()

{
    float matarr1[8][8];/* First matrix store (rowxcolumn)*/
    float matarr2[8][8];/* Second matrix store (rowxcolumn)*/
    float matsum[8][8];/* Sum of matrices store (rowxcolumn)*/
    int i,j,k,l;

    printf("enter order of the two matrices (max 8 rows max 8
    columns) \n");
    scanf("%d %d", &k, &l);
    if(k>8 || l>8)
    {
        printf("error - max of 8 for rows or columns\n");

    }

    else
    {
        printf("enter first matrix\n");
        for(i=0;i<k;i++)
        {
            for(j=0;j<l;j++)
            {
                scanf("%f",&matarr1[i][j]);
            }
        }
```

```c
printf("Your first matrix is \n");
for(i=0;i<k;i++)
{
      for(j=0;j<l;j++)
      {
            printf("%f ",matarr1[i][j]);/* first matrix in
            matarr1 */
      }
      printf("\n");
}
printf("enter second matrix\n");
for(i=0;i<k;i++)
{
      for(j=0;j<l;j++)
      {
            scanf("%f",&matarr2[i][j]);
      }
}
printf("Your second matrix is \n");
for(i=0;i<k;i++)
{
      for(j=0;j<l;j++)
      {
            printf("%f ",matarr2[i][j]);/* second matrix
            in matarr2 */
      }
      printf("\n");
}
/* add correspoding elements of the matrices into matsum */
for(i=0;i<k;i++)
{
      for(j=0;j<l;j++)
      {
            matsum[i][j] = matarr1[i][j] + matarr2[i][j];
      }
}
```

```
            printf("Your matrix sum is \n");
            for(i=0;i<k;i++)
            {
                    for(j=0;j<l;j++)
                    {
                            printf("%f ",matsum[i][j]);/* sum of matrices
                            in matsum */
                    }
                    printf("\n");

            }
      }

   }
```

2. a)

$$\begin{pmatrix} 1.7 & 1.0 & 2.1 \\ 3.1 & 0.4 & 0.0 \\ 3.1 & 6.2 & 2.3 \end{pmatrix}$$

2. b)

$$\begin{pmatrix} 1.3 & 1.0 & -1.3 \\ -3.1 & 0.4 & -0.8 \\ -3.1 & 6.2 & 0.3 \end{pmatrix}$$

3. Code should be as follows.

```
/* Matrix program */
/* multiply two floating point matrices */
#define _CRT_SECURE_NO_WARNINGS
#include<stdio.h>

    int main()

    {
    float matarr1[8][8];/* First matrix store (rowxcolumn)*/
    float matarr2[8][8];/* second matrix store (rowxcolumn)*/
```

```c
float matmult[8][8];/* matrix answer (rowxcolumn)*/
int i,j,k;
int r1,c1,r2,c2;
int error;

error=0;

printf("enter order of the first matrix (max 8 rows max 8
columns) \n");
scanf("%d %d", &r1, &c1);
if(r1>8 || c1>8)
{
    printf("error - max of 8 for rows or columns\n");
    error=1;

}
printf("enter order of the second matrix (max 8 rows max 8
columns) \n");
scanf("%d %d", &r2, &c2);
if(r2>8 || c2>8)
{
    printf("error - max of 8 for rows or columns\n");
    error=1;

}
if(error == 0)

{
    for(i=0;i<r1;i++)
    {
        for(j=0;j<c2;j++)
        {
            matmult[i][j]=0;
        }
    }
```

```c
            printf("enter first matrix\n");
            for(i=0;i<r1;i++)
            {
                 for(j=0;j<c1;j++)
                 {
                       scanf("%f",&matarr1[i][j]);
                 }
            }
            printf("Your first matrix is \n");
            for(i=0;i<r1;i++)
            {
                 for(j=0;j<c1;j++)
                 {
                       printf("%f ",matarr1[i][j]);/* first matrix in
                       matarr1 */
                 }
                 printf("\n");
            }
            printf("enter second matrix\n");
            for(i=0;i<r2;i++)
            {
                 for(j=0;j<c2;j++)
                 {
                       scanf("%f",&matarr2[i][j]);
                 }
            }
            printf("Your second matrix is \n");
            for(i=0;i<r2;i++)
            {
                 for(j=0;j<c2;j++)
                 {
                       printf("%f ",matarr2[i][j]);/* second matrix
                       in matarr2 */
                 }
                 printf("\n");
            }
```

```
/* multiply correspoding elements of the matrices into
matmult */
for(i=0;i<r1;i++)
{
      for(j=0;j<c2;j++)
      {
            for(k=0;k<r2;k++)
            {
                  matmult[i][j] = matmult[i][j] +
                  matarr1[i][k] * matarr2[k][j];
            }
      }
}

printf("Your matrix multiplication is \n");
for(i=0;i<r1;i++)
{
      for(j=0;j<c2;j++)
      {
            printf("%f ",matmult[i][j]);
      }
      printf("\n");

}
  }

}
```

4.

$$\begin{pmatrix} 0.45 & 4.88 & 2.81 \\ 0.57 & 1.85 & 1.04 \\ 2.62 & 5.57 & 2.54 \end{pmatrix}$$

5.

$$\begin{pmatrix} 1 & 0 \\ 0 & 1 \end{pmatrix}$$

CHAPTER 6

1. The code for this is as follows.

```
/*      regression */
/*      user enters points.*/
/*      regression of x on y calculated */
#define _CRT_SECURE_NO_WARNINGS
#include <stdio.h>
#include <math.h>
main()
{

      float xpoints[10], ypoints[10];
      float sigmax, sigmay, sigmaxy, sigmaysquared, xbar, ybar;
      float fltcnt, sxy, syy, c, d;
      int i, points;

      /* User asked for number of points on scatter graph */
      printf("enter number of points (max 10 ) \n");
      scanf("%d", &points);
      if (points > 10)
      {
            printf("error - max of 10 points\n");

      }
      else
      {
            sigmax = 0;
            sigmay = 0;
            sigmaxy = 0;

            sigmaysquared = 0;

            /* User enters points  */
            for (i = 0;i < points;i++)
```

```
{
      printf("enter point (x and y separated by space) \n");
      scanf("%f %f", &xpoints[i], &ypoints[i]);
      sigmax = sigmax + xpoints[i];
      sigmay = sigmay + ypoints[i];
      sigmaxy = sigmaxy + xpoints[i] * ypoints[i];

      sigmaysquared = sigmaysquared + (float)
      pow(ypoints[i], 2);
}
printf("points are \n");
for (i = 0;i < points;i++)
{
      printf(" \n");
      printf("%f %f", xpoints[i], ypoints[i]);

}
printf(" \n");
fltcnt = (float)points;

/* regression variables calculated */
xbar = sigmax / fltcnt;
ybar = sigmay / fltcnt;
sxy = (1 / fltcnt)*sigmaxy - xbar * ybar;

syy = (1 / fltcnt)*sigmaysquared - ybar * ybar;

d = sxy / syy;
c = xbar - d * ybar;

/* Regression line */
printf("Equation of regression line x on y  is\n ");
printf(" x=%f + %fy", c, d);
   }

}
```

2.   The graph is as follows.

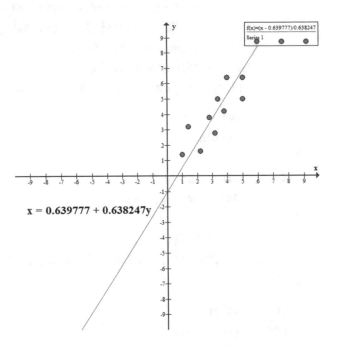

$$x = 0.639777 + 0.638247y$$

3.   PMCC should be −1.

CHAPTER 7

1.   The code and associated graph are as follows.

```
/* simple random walk simulation in 1 dimension */
#define _CRT_SECURE_NO_WARNINGS
#include <stdio.h>
#include <math.h>
#include <stdlib.h>
#include <time.h>

FILE *output;
time_t  t;
```

```c
main()
{
    int i;
    double xrand;

    double x,randwalkarr[20001];

    output= fopen ("randwalk6.dat", "w");    /* external file name */

    for (i=0; i<=20000; i++)
        randwalkarr [i]=0.0;        /* clear array */

    srand((unsigned) time(&t));        /* set the number generator */

      x=0.0;

      for (i=1;i<=20000; i++)
        {
        /* generate x random number */
        xrand=rand()%1000;
        xrand=xrand/1000;
        if(xrand<0.5)
              x=x+1.0;
        else
              x=x-1.0;

          randwalkarr[i] = sqrt(x*x);/* store randwalkarr to total */
    }
    /* Write values to file */
    for (i=0; i<=200; i++)
    {

        fprintf(output,"%d %lf\n", i, randwalkarr[i*100]);
    }

    fclose (output);
}
```

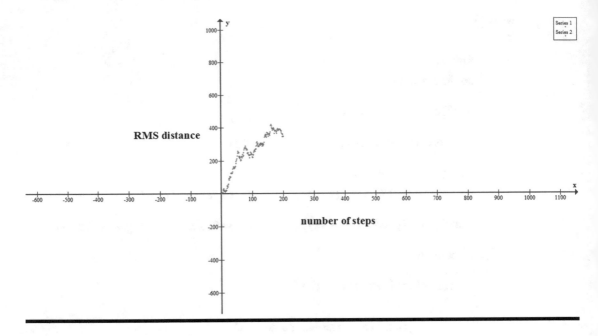

CHAPTER 8

1. a = 1.6   b = 2.8   c = 5.4

2. a = 1.1   b = 2.2   c = 3.3

3. a = 1.1   b = 2.2   c = 3.3   d = 4.4   e = 5.5   f = 6.6

CHAPTER 9

1. The patients' names are Jones, Smith, Stone, Weeks, and Owen. The percentage
    required is 29.411764%. The code for this question is given as follows.

```
/* filereadexr */
/* reads from file */
/* reads and prints sequentially */
/* reads and prints specific records */
/* Does not use seek */
#define _CRT_SECURE_NO_WARNINGS
```

```c
#include<stdio.h>

struct Patient {
    int PatientID;
    char name[13];
    int BloodPressure;
    char allergies;
    char leukaemia;
    char anaemia;
    char asthma;
    char epilepsy;
    char famepil;
};

int main()
{
    FILE *fp;

    struct Patient s2;

    int numread, i;
    double casthepi;
    double percent;

    /* Open patients file */

    fp = fopen("patientex.bin", "r");
    if(!fp)
    {
        printf("patientex.bin file unavailable");
        return(0);
    }
    for (i = 0;i < 17;i++)
    {
        /* Read each patient data from file sequentially */
        fread(&s2, sizeof(s2), 1, fp);
        /* Print patient ID, name and Blood Pressure for each
        patient */

        printf("\nPatientID : %d", s2.PatientID);
        printf("\n Name : %s", s2.name);
```

```c
            printf("\nBloodPressure : %d", s2.BloodPressure);
            printf("\nAllergies %c leukaemia %c anaemia %c",
            s2.allergies, s2.leukaemia, s2.anaemia);
            printf("\nAsthma %c epilepsy %c famely epilepsy %c",
            s2.asthma, s2.epilepsy, s2.famepil);

    }

    fclose(fp);

    /* Re-open the patients file */

    fp = fopen("patientex.bin", "r");
    for (i = 0;i < 17;i++)
    {
        /* Search the file for patient with ID of 23 */

        fread(&s2, sizeof(s2), 1, fp);
        if (s2.PatientID == 23)
        {
            /* Found the patient. Print their name */
            printf("\nName : %s", s2.name);
            break;
        }
    }
    /* Go back to the beginning of the file */

    rewind(fp);
    /* Find all patients with Blood Pressure reading above 63 */

    for (i = 0;i < 17;i++)
    {
        fread(&s2, sizeof(s2), 1, fp);
        if (s2.BloodPressure > 63)
        {
            /* Print out name of each patient with Blood
            pressure above 63 */
            printf("\nName : %s", s2.name);

        }
    }
```

```c
/* Go back to the beginning of the file */
rewind(fp);

/* Read and print out the first 3 patients in the file */

numread = fread(&s2, sizeof(s2), 1, fp);
if (numread == 1)
{
    printf("\nPatientID : %d", s2.PatientID);
    printf("\nName : %s", s2.name);
    printf("\nBloodPressure : %d", s2.BloodPressure);

}
numread = fread(&s2, sizeof(s2), 1, fp);
if (numread == 1)
{

    printf("\nPatientID : %d", s2.PatientID);
    printf("\nName : %s", s2.name);
    printf("\nBloodPressure : %d", s2.BloodPressure);
}
numread = fread(&s2, sizeof(s2), 1, fp);
if (numread == 1)
{
    printf("\nPatientID : %d", s2.PatientID);
    printf("\nName : %s", s2.name);
    printf("\nBloodPressure : %d", s2.BloodPressure);
}
/* Close the file */

fclose(fp);
/* Re-open the patients file */
casthepi = 0;
fp = fopen("patientex.bin", "r");
for (i = 0;i < 17;i++)
{
    /* Search the file for link between asthma and epilepsy */

    fread(&s2, sizeof(s2), 1, fp);
    if (s2.epilepsy == 'y' && s2.asthma == 'y')
```

```
            {
                    casthepi = casthepi + 1.0;
                    /* Found the patient. Print their name */
                    printf("\nLink between asthma and epilepsy");
                    printf("\nName : %s", s2.name);

            }
        }
        percent = (casthepi / 17.0)*100.0;
        printf("\npercent asthma & epilepsy : %f", percent) ;
        fclose(fp);

}
```

2. The two names of the companies asked for are Allenby and Evans LLC. The
    percentage is 11.764706%. The code is as follows.

```
/* filereadex3r */
/* reads from Company file */
/* reads and prints sequentially */
/* reads and prints specific records */
/* does not use seek */
#define _CRT_SECURE_NO_WARNINGS
#include<stdio.h>

struct Company {
        int CompanyID;
        char companyname[13];
        float salesprofitpct;/* profit as a % of sales */
        float totalctrypop;/* total populations countries for sales (in
        millions) */
        float advertpct;/* Advertising as a % of sales */
        float salprofpct;/* Total salaries as a % of profit */
        float mwpct;/* Women as a % of total workers */
        float alienwpct;/* Foreign workers as a % of total */

};
```

```c
int main()
{
    FILE *fp;

    struct Company s2;

    int numread, i;
    double count; /* count of women to men >40 salespercent >40 */
    double percent;/* percent of women to men >40 salespercent >40 */

    /* Open patients file */

    fp = fopen("Companyex.bin", "r");
    if(!fp)
    {
        printf("Companyex.bin file not available");
        return(0);
    }
    for (i = 0;i < 17;i++)
    {
        /* Read and print each Company data from file sequentially */

        fread(&s2, sizeof(s2), 1, fp);
        /* Print Company ID, name etc */
        printf("\nCompanyID : %d", s2.CompanyID);
        printf("\ncompanyname : %s", s2.companyname);
        printf("\nprofit as a percentage of sales : %f",
        s2.salesprofitpct);
        printf("\ntotal populations countries for sales (in
        millions) %f ", s2.totalctrypop);
        printf("\nAdvertising as a percentage of sales %f ",
        s2.advertpct);
        printf("\nTotal salaries as a percentage of profit %f ",
        s2.salprofpct);
        printf("\nWomen as a percentage of total workers %f ",
        s2.mwpct);
        printf("\nForeign workers as a percentage of total %f ",
        s2.alienwpct);

    }
```

```
        fclose(fp);

        /* Re-open the Company file */
        fp = fopen("Companyex.bin", "r");
        for (i = 0;i < 17;i++)
        {
            /* Search the file for Company with ID of 23 */

            fread(&s2, sizeof(s2), 1, fp);
            if (s2.CompanyID == 23)
            {
                /* Found the company. Print their name */
                printf("\nCompany with ID of 23 ");
                printf("\nCompany Name : %s", s2.companyname);
                break;
            }
        }
        /* Go back to the beginning of the file */

        rewind(fp);
        /* Find all Companys with women to men percent > 50  */

        for (i = 0;i < 17;i++)
        {
            fread(&s2, sizeof(s2), 1, fp);
            if (s2.mwpct > 50)
            {
                /* Print out name of each company with women to men
                percent > 50*/
                printf("\nwomen to men >50pc");
                printf("\nCompany Name : %s", s2.companyname);

            }
        }
        /* Go back to the beginning of the file */

        rewind(fp);

        /* Read and print out the first 3 Companys in the file */

        numread = fread(&s2, sizeof(s2), 1, fp);
        printf("\nFirst 3 companies on file");
```

```c
if (numread == 1)
{
     printf("\nCompanyID : %d", s2.CompanyID);
     printf("\nCompany Name : %s", s2.companyname);

}
numread = fread(&s2, sizeof(s2), 1, fp);
if (numread == 1)
{

     printf("\nCompanyID : %d", s2.CompanyID);
     printf("\nCompany Name : %s", s2.companyname);

}
numread = fread(&s2, sizeof(s2), 1, fp);
if (numread == 1)
{

     printf("\nCompanyID : %d", s2.CompanyID);
     printf("\nCompany Name : %s", s2.companyname);

}
/* Close the file */

fclose(fp);
/* Re-open the patients file */
count = 0;/* set count of percent women to men >40
salespercent >40 */
fp = fopen("Companyex.bin", "r");
for (i = 0;i < 17;i++)
{
     /* Search count of percent women to men >40
     salespercent >40 */

     fread(&s2, sizeof(s2), 1, fp);
     if (s2.mwpct > 40.0 && s2.salesprofitpct > 40.0)
     {
          count = count + 1.0; /* Add 1 to overall count */

          /* Found the company. Print their name */

          printf("\nLink between women to men >40pc and
          salespc > 40 ");
```

```
                    printf("\nName : %s", s2.companyname);

        }

    }
    /* Calculate and print percentage of women to men over 40 and
    salespercent over 40 */
    percent = (count / 17.0)*100.0;
    printf("\npercent women to men >40 salespercent >40 : %f", percent);

    fclose(fp);
}
```

CHAPTER 10

1. x = 5.099998,   y = 0.146350

2. x = 2.00000,   y = 1.654713

3. x = 1.00000,   y = 1.6966

4. x = 1.004999   y = 099075

The curve for the output data is shown in red and compared with the correct curve.

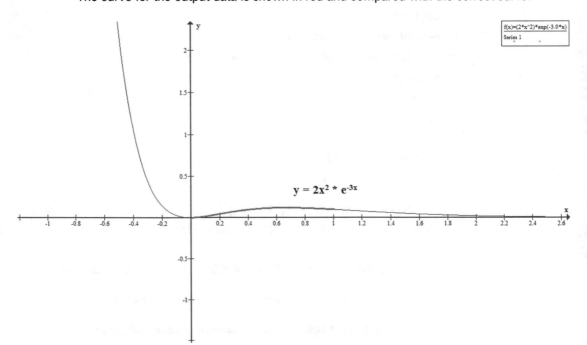

5.   x = 1.000000   y = 1.978569

The curve for the output data is shown in red and compared with the
correct curve.

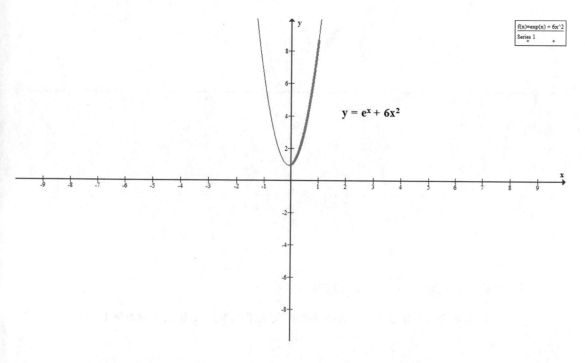

$$y = e^x + 6x^2$$

6.   x = 2.005000   y = 19.958468

The curve for the output data is shown in red and compared with the
correct curve.

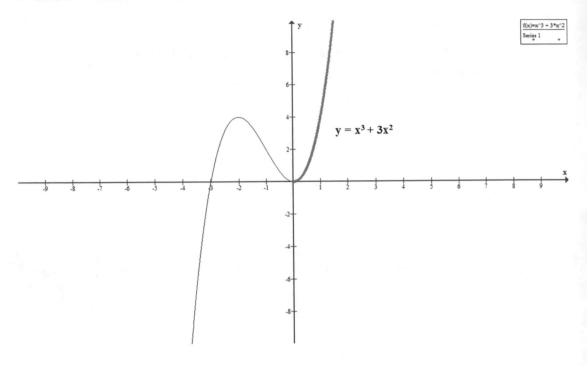

$$y = x^3 + 3x^2$$

7.   x = 1.000000   y = 0.096944

The curve for the output data is shown in red and compared with the correct curve.

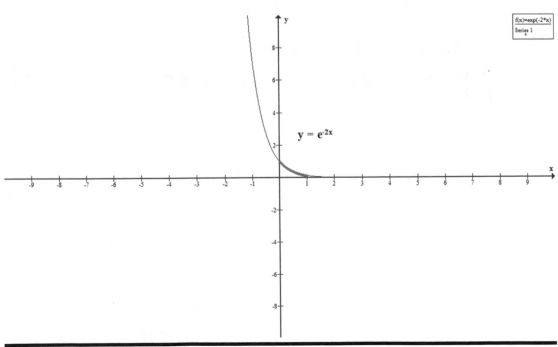

$$y = e^{-2x}$$

Index

A, B

Add, two numbers
   code, 4
   %d, 5
   int, 4
   scanf, 5
   total, 5
Augmented matrix program
   funcdivide and funcsubtract
      forloops, 175, 176
      functions, 173, 175
      INF error, 181
      matrix, 177, 179
      procedures, 171
      swopping mechanism, 183–189
      testing, 180, 181, 190
   matrix[3][4] array, 162
   nine-stage mechanism, 182, 183
      stage 1-3, divide, 163–165
      stage 4-5, subtract, 165–166
      stage 6-7, divide, 167–168
      stage 8, subtract, 169
      stage 9, divide, 169, 170

C

char c, 3
Command line environment, 256, 257
Company records file
   business practice, 229
   Companyex.bin, 229
   ID, 230
   structure, 230–232
Compiling, 2, 63
Completing the square method, 45, 46

D

Data arrays
   char arr, 19, 20
   int array, 19, 20
   reads data, 21, 22
   2D array, 22–24
data_record.matrix, 210, 212
Data vetting, 22, 36
Debugging, 252
Decimal numbers, add, 5
Differential equations
   algebraic function, 235
   first order, 246
   integral calculus, 235
   second order, 246–249
Divide two numbers, 7
Do loop, 13, 266
Double, float, and integer, compare, 263

E

Euler method
   analysis, 237
   continuous blue curve, 241
   curve gradient, 237

© Philip Joyce 2019
P. Joyce, *Numerical C*, https://doi.org/10.1007/978-1-4842-5064-8